Adipose Tissue Development
From Animal Models to Clinical Conditions

Endocrine Development

Vol. 19

Series Editor

P.-E. Mullis Bern

3rd ESPE Advanced Seminar in Developmental Endocrinology, Paris, March 12–13, 2009

Adipose Tissue Development

From Animal Models to Clinical Conditions

Volume Editors

Claire Levy-Marchal Paris
Luc Pénicaud Dijon

11 figures and 3 tables, 2010

Basel · Freiburg · Paris · London · New York · Bangalore · Bangkok · Shanghai · Singapore · Tokyo · Sydney

Endocrine Development

Founded 1999 by Martin O. Savage, London

Claire Levy-Marchal
Hôpital Robert Debré
Unité INSERM 690
Paris, France

Luc Pénicaud
Centre des Sciences du Goût et de l'Alimentation
UMR 6265 CNRS, 1324 INRA-uB
Dijon, France

Library of Congress Cataloging-in-Publication Data

ESPE Advanced Seminar in Developmental Endocrinology (3rd : Paris, France :
2009)
  Adipose tissue development : from animal models to clinical conditions /
3rd ESPE Advanced Seminar in Developmental Endocrinology, Paris, March
12-13, 2009 ; volume editors, Claire Levy-Marchal, Luc Pénicaud.
    p. ; cm. -- (Endocrine development, ISSN 1421-7082 ; v. 19)
  Includes bibliographical references and index.
  ISBN 978-3-8055-9450-9 (hard cover : alk. paper)
  1. Adipose tissues--Pathophysiology--Congresses. 2. Adipose
tissues--Diseases--Endocrine aspects--Congresses. 3. Metabolic
syndrome--Congresses. 4. Developmental endocrinology--Congresses. I.
Levy-Marchal, C. (Claire) II. Pénicaud, Luc. III. Title. IV. Series:
Endocrine development, v. 19. 1421-7082 ;
  [DNLM: 1. Adipose Tissue--physiology--Congresses. 2.
Adipocytes--pathology--Congresses. 3. Adipose
Tissue--physiopathology--Congresses. QS 532.5.A3 E77a 2010]
  RB147..E87 2010
  571.5'7--dc22

                                        2010010545

Bibliographic Indices. This publication is listed in bibliographic services, including Current Contents® and PubMed/
MEDLINE.

© Copyright 2010 by S. Karger AG, P.O. Box, CH–4009 Basel (Switzerland)
www.karger.com
Printed in Switzerland on acid-free and non-aging paper (ISO 9706) by Reinhardt Druck, Basel
ISSN 1421–7082
ISBN 978–3–8055–9450–9
e-ISBN 978–3–8055–9451–6

Contents

Preface

Adipose tissue has been looked at with a new interest for the past years. Previously seen as a storage organ involved only in fuel metabolism, it is now regarded as an actual endocrine organ. The discovery of a number of adipokines secreted by adipose tissue and involved in the regulation of energy balance, fuel and lipid metabolism, and insulin sensitivity makes it an organ of major interest for new physiological concepts and a major target in the prevention and treatment of a number of clinical conditions. However, little is known today with respect to the interplay between adipocytes and the stromal components of adipose tissue, not only in terms of physiology in the mature tissue, but also in terms of development.

Thanks to the endeavor of the European Society for Pediatric Endocrinology, a seminar dedicated to junior physicians and scientists was held in Paris in March 2009 on the topic of development of adipose tissue. This seminar gathered about 35 young members of the Society from all over Europe to listen to and debate with distinguished international investigators and scientists in the field.

This book encompasses the proceedings of the conferences covering basic knowledge and approaches as well as clinical investigations and experiences.

Adipocytes arise from mesenchymal stem cells by a sequential pathway of differentiation. White adipocytes differentiate from various types of vascular cell types, probably located within the white adipose tissue itself. Brown adipocytes arise from myogenic precursors. The differentiation between white adipocyte and brown adipocyte lineages occurs in the earliest steps of fetal development, and both phenotypes are acquired independently. A better knowledge of these differentiation pathways is crucial for the development, among others, of new drugs in the fight against obesity and the metabolic consequences.

These proceedings cover the importance of nervous regulation of both white and brown adipose tissue mass with a review of the different physiological parameters which are regulated such as metabolism (lipolysis and thermogeneis) and secretory activity (leptin and other adipokines), but also the plasticity of adipose tissues (proliferation differentiation and apoptosis) showing the presence of a neural feedback loop between adipose tissues and the brain which plays a major role in the regulation of energy homeostasis.

The discovery of leptin has clearly demonstrated a relationship between body fat and the neuroendocrine axis since leptin influences appetite and the reproductive axis. Since adipose tissue is a primary source of leptin, adipose tissue is no longer considered as simply a depot to store fat. Recent findings demonstrate that numerous other genes, i.e. neuropeptides, interleukins and other cytokines, and biologically active substances such as leptin and insulin-like growth factors I and II are also produced by adipose tissue, which could influence appetite and the reproductive axis. Targets of leptin in the hypothalamus include neuropeptide Y, proopiomelanocortin and kisspeptin. These few lines depict the complexity of the cross-talk between the brain and adipose tissue as far as the reproductive function is concerned.

A more recent observation is the relation between obesity and cancer. In addition to diabetes and cardiovascular diseases, epidemiological evidence demonstrates that people who are obese or overweight are at increased risk of developing cancer – colon, breast (in postmenopausal women), endometrial or kidney cancer being among the most frequent. In addition to the increase in tumor occurrence, obesity also affects tumor prognosis, especially in breast and prostate cancers. In breast cancer, obesity is associated with reduced survival and increased recurrence independent of menopausal status. Host factors seem to contribute to the occurrence of tumors exhibiting an aggressive biology defined by advanced stages and high grade. Mature adipocytes are part of the breast cancer tissue and as highly endocrine cells are susceptible to profoundly modify breast cancer cell behavior.

It was demonstrated more than 10 years ago that the development of obesity is determined as early as during fetal life and early infancy. The epidemiological evidence is reviewed here. Early puberty and age at menarche are consequences of rapid infant weight gain and childhood overweight, and in turn these adolescent traits are predictive for obesity, diabetes, hypertension and cardiovascular disease events in later life. An understanding of the nutritional, parental and wider determinants of rapid infant weight gain is important for the development of obesity prevention strategies starting in early life.

A clinical model of the development of fat mass early in life following fetal growth restriction is proposed with respect to the development of insulin resistance and to the metabolic syndrome. Over the last 15 years a number of long-term health risks associated with reduced fetal growth have been identified, including cardiovascular diseases, hypertension, dyslipidemia, and type 2 diabetes. A common feature of these conditions is insulin resistance, which is thought to play a pathogenic role. However, despite abundant data in the literature, it is still difficult to trace the pathway by which fetal events, environmental or not, may lead to the increased morbidity later in life. To explain this association, several hypotheses have been proposed pointing to the role of either a detrimental fetal environment or a genetic susceptibility or an interaction between the two and of the particular dynamic changes in adiposity that occur during catch-up growth.

The metabolic syndrome defines the clustering of cardiovascular risk factors and is driven by peripheral insulin resistance. The 'driving force' of the syndrome, i.e. insulin resistance, develops mainly in obese children due to a specific pattern of lipid partitioning characterized by increased deposition of fat in the visceral compartment as well as in insulin-responsive tissues, such as muscle and liver. Such a lipid deposition pattern results in peripheral insulin resistance and a compensatory hyperinsulinemia. The definition of the syndrome in childhood suffers from many limitations related to different ethnic characteristics as well as age and development dependency of some of the components. Despite these limitations, the clustering of risk factors characteristic of the syndrome in childhood is associated with accelerated atherogenesis in adulthood. These complications are one of the major future concerns of public health with the rising incidence of overweight and obesity in the youth.

Human lipodystrophies represent a heterogeneous group of diseases characterized by generalized or partial fat loss, with fat hypertrophy in other depots when partial. Insulin resistance, dyslipidemia and diabetes are generally associated with leading to early complications. Whereas genetic forms are rare and represent a unique clinical model for the development of adipose tissue, acquired forms are often iatrogenic.

This splendid collection of investigation data and reviews will with no doubt serve as a reference for all pediatricians and scientists interested by obesity, endocrinology and development.

Claire Levy-Marchal, Paris

Levy-Marchal C, Pénicaud L (eds): Adipose Tissue Development: From Animal Models to Clinical Conditions.
Endocr Dev. Basel, Karger, 2010, vol 19, pp 1–20

Human Lipodystrophies: Genetic and Acquired Diseases of Adipose Tissue

Jacqueline Capeau[a–c] · Jocelyne Magré[a,b] ·
Martine Caron-Debarle[a,b] · Claire Lagathu[a,b] ·
Bénédicte Antoine[a,b] · Véronique Béréziat[a,b] · Olivier Lascols[a–c] ·
Jean-Philippe Bastard[a–c] · Corinne Vigouroux[a–c]

[a]INSERM, U938, CDR Saint-Antoine; [b]UPMC University Paris 06, UMR_S938 and [c]AP-HP Tenon and Saint-Antoine Hospitals, Paris, France

Abstract

Human lipodystrophies represent a heterogeneous group of diseases characterized by generalized or partial fat loss, with fat hypertrophy in other depots when partial. Insulin resistance, dyslipidemia and diabetes are generally associated, leading to early complications. Genetic forms are uncommon: recessive generalized congenital lipodystrophies result in most cases from mutations in the genes encoding seipin or the 1-acyl-glycerol-3-phosphate-acyltransferase 2 (AGPAT2). Dominant partial familial lipodystrophies result from mutations in genes encoding the nuclear protein lamin A/C or the adipose transcription factor PPARγ. Importantly, lamin A/C mutations are also responsible for metabolic laminopathies, resembling the metabolic syndrome and progeria, a syndrome of premature aging. A number of lipodystrophic patients remain undiagnosed at the genetic level. Acquired lipodystrophy can be generalized, resembling congenital forms, or partial, as the Barraquer-Simons syndrome, with loss of fat in the upper part of the body contrasting with accumulation in the lower part. Although their etiology is generally unknown, they could be associated with signs of autoimmunity. The most common forms of lipodystrophies are iatrogenic. In human immunodeficiency virus-infected patients, some first-generation antiretroviral drugs were strongly related with peripheral lipoatrophy and metabolic alterations. Partial lipodystrophy also characterize patients with endogenous or exogenous long-term corticoid excess. Treatment of fat redistribution can sometimes benefit from plastic surgery. Lipid and glucose alterations are difficult to control leading to early occurrence of diabetic, cardiovascular and hepatic complications.

Diseases of adipose tissue are present with a high prevalence in the global population, in particular those linked with fat expansion leading to obesity, metabolic syndrome or type 2 diabetes. The consequences of increased fat depots are markedly dependent upon their localization. Adipose tissue in the lower part of the body is able to expand and can therefore accumulate excessive energy from diet, stored as triglycerides: taken as a whole, it appears protective at the metabolic level [1]. By contrast, accumulation of fat in the upper part of the body is deleterious. Most abdominal fat is accounted for by subcutaneous fat (SAT) and, under physiologic conditions, only a minor part is represented by intra-abdominal visceral fat (VAT). Excessive SAT, and even more VAT, is strongly associated with metabolic alterations and insulin resistance. These alterations result from the increased release of free fatty acids (FFA) from insulin-resistant adipocytes but also from modified adipokine production with decreased production of adiponectin by adipocytes and increased production of pro-inflammatory cytokines (IL-6, IL-1β, TNF-β) and chemokines (such as CCL2), in part by adipocytes, but mainly by macrophages invading adipose tissue [2].

Human lipodystrophies are far less common than obesity and characterized, at the opposite, by fat disappearance. However, at the metabolic level, common alterations are observed with insulin resistance, dyslipidemia and generally increased FFA and decreased adiponectin production. Recently, the genetic origin of some of these lipodystrophies has been clarified. However, a number of patients remain undiagnosed at the genetic level. Otherwise, some acquired forms are iatrogenic [3–6].

Definition and Diagnosis

Human lipodystrophies represent a heterogeneous group of diseases [4–6] characterized by the loss of body fat, which could be localized or generalized. If localized, it is often associated with fat hypertrophy in some other depots.

At the clinical level, peripheral lipoatrophy affecting SAT can be easily diagnosed when marked and affecting regions with a natural large fat thickness. Loss of fat into cheeks and temples gives a gaunt face and, in the limbs, makes muscles and veins highly visible. However, lipoatrophy can be difficult to diagnose if mild, in particular at the lower limb level in males, who can have physiologically a low fat amount. In those cases, a CT scan at the thigh level is useful but requires comparisons with normal subjects. The diagnosis of visceral fat atrophy (or hypertrophy) requires imaging techniques: a CT scan or a MRI at the lumbar L4 level allows precise evaluation of the SAT and VAT areas.

Human lipodystrophies are generally associated with severe insulin resistance. Therefore, clinical signs of insulin resistance can help diagnosis: the presence of skin lesions of acanthosis nigricans, a skin-brownish lesion present in the axillae, neck and other body folds is an excellent indication of marked insulin resistance, in particular in normal-weight patients. Long-term insulin resistance can lead to acromegaloid features, striking in face and seen in particular in congenital forms. Insulin resistance can result in increased size of genital organs in prepubertal children, ovarian hyperandrogenism leading to virilization and hirsutism with polycystic ovary syndrome and hyperthecosis in women. Insulin resistance is also commonly associated with hepatomegaly and steatosis.

At the metabolic level, lipodystrophies are characterized by glucose and lipid alterations which can be mild or even absent during childhood and increase in severity when patients age. Lipid alterations associate increased triglyceride (TG) level, which can be raised up to 100 mmol/l, leading to a high risk of acute pancreatitis, while HDL cholesterol is decreased. Glucose values could remain in the normal range in young patients, if insulin secretion is able to compensate for insulin resistance, but increase progressively leading to glucose intolerance then diabetes, difficult to control.

The main acute complication is acute pancreatitis due to very high TG level. Chronic complications are related to long-term diabetic complications, microangiopathy, affecting retina, kidney and nerves, macroangiopathy, leading to early atherosclerosis, and to hepatic complications of steatosis leading to steatohepatitis and sometimes cirrhosis and portal hypertension.

The differential diagnosis with syndromes of insulin resistance due to alterations at the insulin receptor level (leprechaunism, type A and type B syndromes) could be difficult. However, in these later syndromes, lipodystrophy and dyslipidemia are absent and very high adiponectin levels have been recently reported [7]. Lipomatosis represents localized fat tumors, different from lipodystrophies. They can be multiple, affecting mainly the proximal limbs areas and the neck in the familial lipomatosis, and are sometimes associated with mutations in mitochondrial DNA (*MERRF* mutations in particular). The Launois-Bensaude lipomatosis, of unknown origin, is often associated with peripheral neuropathy and increased alcohol intake.

Classification

Human lipodystrophies can be defined by the extent of fat loss (generalized or partial) and by their etiology, either genetic or acquired (table 1). Genetic forms of lipodystrophy are uncommon diseases and, up to now, only a few genes have been identified, the alteration of which is responsible for lipodystrophy and insulin resistance.

Genetic forms of complete lipodystrophy called Berardinelli-Seip congenital lipodystrophy (BSCL) or congenital generalized lipodystrophy are exceptional, with fat loss being generally recognized at birth or very early in infancy. It is associated

Table 1. Classification and main clinical features of lipodystrophies

		Trans-mission	Protein encoded by the altered gene or causal agent	Age at onset lipody-strophy	Adipose distribution	Clinical and biological parameters
Genetic	Generalized					
	BSCL1	AR	AGPAT2	Birth or early infancy	Complete lipoatrophy	Acanthosis nigricans Dyslipidemia Diabetes
	BSCL2	AR	Seipin			
	BSCL3	AR	Caveolin 1			
	Partial					
	FPLD2	AD	Lamin A/C	Puberty	Limbs and buttocks lipoatrophy, increased fat in the face and neck	
	Metabolic lamino-pathy	Gen-erally AD	Lamin A/C	Puberty	Mild or absent lipodystrophy	Acanthosis nigricans Dyslipidemia Diabetes
	Partial					
	FPLD3	AD	PPARγ	Puberty	Lower body lipoatrophy	Hypertension Acanthosis nigricans Dyslipidemia Diabetes
	Partial					
	AKT2-linked	AD	AKT2/PKB		Partial lipodystrophy	Hypertension Acanthosis nigricans Diabetes
Acquired	Generalized					
	Lawrence syndrome		Unknown Sometimes autoimmune disorders	Childhood or adulthood	Complete lipoatrophy	Sometimes panniculitis Acanthosis nigricans Dyslipidemia Diabetes
	Partial					
	Barraquer-Simmons syndrome		Unknown Lamin B2 proposed but not confirmed	Adolescence or early adulthood	Upper body lipoatrophy Lower body fat accumulation	Uncommon metabolic alterations Sometimes low C3 and membrano-proliferative glomerulonephritis
	Generalized or partial					
	HIV-related		Some antiretro-viral drugs: stavudine, zidovudine, first-generation protease inhibitors	Generally adulthood	Peripheral lipoatrophy Central lipoatrophy or fat accumulation	Dyslipidemia Sometimes diabetes

Capeau · Magré · Caron-Debarle · Lagathu · Antoine · Béréziat · Lascols · Bastard · Vigouroux

	Trans-mission	Protein encoded by the altered gene or causal agent	Age at onset lipody-strophy	Adipose distribution	Clinical and biological parameters
Partial					
Related to hyper-corticism		Endogenous or exogenous excess cortisol	Generally adulthood	Lower body lipoatrophy Upper body fat accumulation	Dyslipidemia Often diabetes

AR = Autosomal recessive, AD = autosomal dominant, BSCL = Berardinelli-Seip congenital lipodystrophy, FPLD = familial partial lipodystophy, HIV = human immunodeficiency virus.

with severe insulin resistance. Most of the patients present recessive mutations in one of two genes, *BSCL2* encoding seipin or 1-acyl-glycerol-3-phosphate-acyltransferase-2 *(AGPAT2)*. A third gene, *CAV1*, encoding caveolin 1, has been recently identified in one patient. At present, less than 5% of patients with congenital generalized lipodystrophy remain without identified genetic alteration.

In partial lipodystrophies, which are rare diseases, two major genes have been identified so far that present generally heterozygous mutations: *LMNA*, encoding lamin A/C and *PPARG*, encoding PPARγ. Mutations in LMNA are more frequent than those in PPARG and can lead to a number of phenotypes, among which a phenotype where severe insulin resistance is the dominant feature, now designed as 'metabolic laminopathy' [8]. In an international effort searching for new disease-causative genes, mutations in *AKT2*, *LMNB2* encoding lamin B2, *CAV1* and CIDEC were reported in a few patients which remain isolated cases. Numerous patients remain undiagnosed at the genetic level.

Those observed in human immunodeficiency virus (HIV)-infected patients and attributed to the antiretroviral treatment mainly represent acquired forms of lipodystrophy. Very recent data suggest that the chronic viral infection could be also involved. New antiretroviral drugs, with less adverse effects on adipose tissue, are now used. Therefore, lipodystrophy is now less prevalent in this population. However, a number of comorbidites related to insulin resistance and aging occur at an early age in these patients.

A number of acquired lipodystrophies have been recognized for a long time in some rare patients. These forms can be either generalized, as the Lawrence syndrome, or partial, as the Barraquer-Simons syndrome. Their origin is unknown even if immune alterations and signs of autoimmunity have been indentified in some patients. Otherwise, patients with hypercortisolism, either endogenous or exogenous, often present fat redistribution with loss of fat in the limbs and buttocks and increased fat in the upper part of the body, and in particular at the back of the neck (buffalo hump).

Finally, fat redistribution with loss of fat in the periphery and increased fat at the central level, is a physiologic evolution during aging. This central fat redistribution is associated with metabolic alterations such as insulin resistance, increased prevalence of diabetes and dyslipidemia. This could represent a very mild and physiologic form of lipodystrophy with associated metabolic abnormalities. This central fat redistribution is exacerbated in the metabolic syndrome with associated metabolic alterations leading to an increased risk of cardiovascular disease and of diabetes.

Pathophysiology of Adipose Tissue Loss

Adipose tissue now appears as playing a leading role in energy metabolism and insulin sensitivity through the control of lipid metabolism and the secretion of numerous adipokines involved in important functions and in particular in insulin sensitivity (mainly adiponectin) and insulin resistance (pro-inflammatory cytokines). When fat depots are reduced due to lipoatrophy, as seen in these patients, TG present on circulating lipoproteins, chylomicrons and VLDL can be only partially stored in fat depots leading to increased circulating TG [9]. In addition, the hydrolysis of TG on lipoproteins, occurring inside the vascular lumen, leads to increased circulating FFA levels. Reduced fat amounts result in reduced circulating leptin [10] levels that are strongly related to the total amount of fat and in particular of SAT. Very low levels of leptin are deleterious for metabolism and leptin replacement therapies were shown to markedly improve metabolic parameters in patients with severe lipodystrophies. Adiponectin values are also generally greatly reduced [10] in association with strong insulin resistance. Adiponectin is important to oxidize FFA into mitochondria in the liver and muscles. Therefore, adiponectin deficiency impairs this oxidation leading to intracellular accumulation of fatty acid derivatives. Increased FFA and decreased adiponectin are two major actors in the process called lipotoxicity related to an ectopic accumulation of TG associated with insulin resistance [11]. The mechanisms postulated for lipotoxicity imply the increased level of fatty acid derivatives, acyl-CoA, diglycerides, ceramides, present in the cytosol of some tissues such as the liver, muscle, heart and pancreas, due to the decreased ability of mitochondria to oxidize acyl-CoA. Accumulation of fatty acid derivatives leads to activation of stress and inflammatory kinases such as IKK-β and some PKC isoforms, which phosphorylate on specific serine residues the insulin receptor substrate protein (IRS1). This

Capeau·Magré·Caron-Debarle·Lagathu·Antoine·Béréziat·Lascols·Bastard·Vigouroux phosphorylation blocks insulin signaling, resulting in decreased glucose transport inside muscles. Excessive fatty acids are derived towards TG, which depot in the cytosol leading to steatosis in the liver and intramyocellular fat deposits in the muscles and heart. This lipid deposition buffers excessive fatty acids derivatives and preserves tissues from further damage. It can be diagnosed by imaging methods as nuclear magnetic resonance spectroscopy. Other pathways could be altered resulting in increased gluconeogenesis and increased hepatic glucose production.

Altogether, this lipid deposit and associated alterations lead to insulin resistance, which could result in altered glucose tolerance then diabetes, when pancreatic insulin secretion is unable to fully compensate for insulin resistance. However, this post-receptor insulin resistance is selective, and the inability of insulin to suppress gluconeogenesis contrasts with the preserved capacity of the hormone to activate de novo lipogenesis through the SREBP-1 pathway. As a result, hyperinsulinemia increases de novo lipid production responsible for increased TG-rich VLDL production and hypertriglyceridemia [9].

Genetic Lipodystrophies

Berardinelli-Seip Congenital Generalized Lipodystrophy
The main genetic form of generalized lipodystrophy is a very rare disease, Berardinelli-Seip congenital lipodystrophy or BSCL, previously denominated lipoatrophic diabetes. Two main forms, BSCL1 and 2, have been described.

At the clinical level, lipoatrophy is neonatal or very early and complete affecting both SAT and VAT. In early infancy, patients present with muscular hypertrophy and organomegaly, in particular cardiac hypertrophy, associated with an increased growth velocity. Metabolic complications generally appear progressively during childhood, with increased TG level, then glycemia leading to overt insulin-resistant diabetes at puberty. Insulin resistance is generally present during childhood, with skin acanthosis nigricans being present, but not always diagnosed if insulin levels are not measured. Early complications, in particular those related to diabetes, occur in adults.

The genetic origin of BSCL2 was identified at first by Magré et al. [12] in 2001 as resulting from mutations in *BSCL2* encoding seipin of unknown function. In most cases, patients have homozygous mutations but can also be compound heterozygotes. Up to now, 31 different mutations have been identified in 136 patients. All, but 3, are null mutations. BSCL2 patients are more severely affected than BSCL1 and often present mild mental retardation. Patients are often Caucasian with a higher prevalence of the disease in Europe, the Middle East and Asia. In addition, patients from Brazil, probably of Portuguese origin, have been identified.

The function of seipin remained unknown up to recently when data obtained with yeast and human cells implicated this protein in lipid metabolism, more specifically in lipid droplet formation. Seipin deficiency results in severe alterations in lipid droplets morphology indicative of a defect in the formation or maturation of this organelle [3, 13]. In cells issued from patients with seipin mutations, this alteration in the pattern of lipid droplets was associated with a decreased activity of the stearoyl-CoA desaturase-1 (SCD1) reflected by an increase ratio of saturated to the corresponding monounsaturated fatty acids in cellular TG and phospholipids [14]. Unsaturated but not saturated fatty acids are able to induce the formation of new and/or increase the size of preexisting lipid droplets. SCD1 plays a key role in this process by partitioning excess lipid into monounsaturated fatty acids that can be safely stored. The precise mechanism by which seipin works with SCD1 at the endoplasmic reticulum level to synthesize lipid droplets requires further investigation.

BSCL1, identified as linked to a locus on chromosome 9q34, was related later on with mutations in *AGPAT2* by the group of Garg et al. [15] in 2002. This gene is mutated in about 50% of patients with typical BSCL and the disease is recessively inherited, most patients being homozygous. Thirty-three different mutations have been described in 110 patients: most are null mutations and 8 are missense mutations. This form is mainly observed in patients of African ancestry but also in Caucasian patients. At the clinical level, lipoatrophy implicates all fat depots except mechanical fat. Metabolic alterations are milder than those reported in BSCL2 patients and mental retardation is usually absent.

AGPAT2, the most expressed AGPAT adipocyte isoform, catalyzes acylation of lysophosphatidic to phosphatidic acid in the pathway of triglycerides synthesis [3]. Accordingly, its reduced level or absence could explain decreased TG accumulation and altered adipocyte differentiation. In addition, the modified levels of lysophosphatidic and phosphatidic acids, which exert important signaling functions, could also play a role.

Apart from these two genes, a homozygous nonsense mutation, Glu38X, in a third gene, *CAV1,* encoding caveolin 1, was recently identified in a Brazilian patient by Magré and coworkers [16]. The phenotype is very similar to that of BSCL1 and BSCL2 patients. Because caveolin 1 plays an important role in the entry of lipids towards intracellular lipid droplets in adipocytes, its absence could explain decreased fat. In addition, a role for caveolin in insulin signaling has been reported which could explain the phenotype of severe insulin resistance.

Taken as a whole, the mutations in these three genes account for 98% of the patients we investigated: among 117 patients, 62 are mutated in the seipin gene, 60 in *AGPAT2*, and 1 in *CAV1*. Only 3 patients remain undiagnosed at the genetic level.

To explain the severe insulin resistance associated with lipoatrophy, it can hypothesized that lipotoxicity is particularly severe in these patients, completely unable to store fat in adipose tissue and with massive ectopic lipid depots in the muscles, heart and liver. In addition, very low leptin levels are involved in severe metabolic alterations and could possibly be reverted by a treatment with recombinant leptin. Decreased levels of adiponectin were also previously reported, at least in patients with mutations in *AGPAT2* and *CAV1* [10, 16].

Therefore, whatever the mechanism leading to fat loss, the almost complete absence of fat in humans is associated with severe insulin resistance and metabolic alterations leading to early complications and reduced lifespan.

Partial Lipodystrophies Linked to LMNA Mutations
Partial lipodystrophies dominantly inherited are often denominated familial partial lipodystrophy or FPLD and two main forms have been identified at the genetic level. At first, mutations in *LMNA* were recognized as responsible for the FPLD of the Dunnigan type (also called FPLD2). Then, in a few patients, mutations in the gene encoding PPARγ were found responsible for FPLD3.

The story of *LMNA* mutations is fascinating. In less than 10 years, ten different diseases were shown to be linked to a number of mutations in the gene encoding lamin A/C: the phenotypes cover a large spectrum, some of them overlapping each other, resulting in a continuum of diseases collectively called laminopathies, affecting in priority tissues of mesenchymal origin. The story began in 1999 by the discovery by G. Bonne working in the group of K. Schwartz in Paris that this gene was responsible for the dominantly inherited Emery-Dreyfuss myopathy [17]. Later on, other forms of myopathies and of cardiomyopathies were found to be related to *LMNA* mutations. In 2000, the Canadian group of Robert Hegele and the English group of Richard Trembath found that the dominantly inherited FPLD of the Dunnigan type was also due to *LMNA* mutations, most of them being located in exon 8, coding for the globular C-terminal domain of the protein, with a hot-spot at residue 482 replacing arginine by a neutral residue in 90% of FPLD2 patients [17]. Further studies on the patients' phenotype were performed by our group [18] and others. Indeed, the typical FPLD2 phenotype due to the R482 *LMNA* substitution is absent in prepubertal children. The clinical and biological signs appear after puberty associating fat loss at the limb and abdominal subcutaneous level together with increased fat in the face and neck giving a cushingoid appearance. The clinical phenotype is generally obvious in women who often complain also of hirsutism, but can be very mild in men, which could make the diagnosis difficult. Metabolic alterations also occur after puberty leading to severe hypertriglyceridemia, insulin-resistant diabetes and early atherosclerosis.

The pathophysiology of the disease is not well understood. Lamin A/C together with lamin B are intermediate filaments present inside the nucleus where they form a meshwork under the inner nuclear membrane, the nuclear lamina. Lamina is associated to the membrane through interactions with different proteins embedded in the inner nuclear membrane that also interacts with other partners across the nuclear envelope, like nesprin, linking the nucleus to the actin cytoskeleton in the cytosol. In addition, some lamin A/C is present inside the nucleoplasm and probably plays important functions at that level: lamin A/C is able to interact with other proteins controlling nuclear functions such as LAP2α, the retinoblastoma protein Rb, cFos controlling the transcription factor AP-1 and SREBP-1 a transcription factor playing an important role in adipocyte and controlling PPARγ.

Mature lamin A is obtained after a complex process of maturation. During this process, prelamin A gains a farnesyl anchor, which allows the movement of the protein to the inner nuclear envelope. The maturation then requires the action of a specific protease called ZMP-STE24 or FACE1, which removes the C-terminal end of the protein including the farnesyl anchor. Thus the link between lamin A and the nuclear membrane is weakened allowing lamin A to be partially localized in the nucleoplasm.

The pathophysiology of the lipodystrophic phenotype observed in FPLD2 patients remains unknown. The C-terminal domain of lamin A/C has been shown to interact with DNA and the transcription factor SREBP-1, thus playing a scaffolding function. When lamin A/C is mutated on the residues responsible for FPLD2, all located at the surface of the globular domain, a decreased interaction with DNA and SREBP-1 has been reported which could explain altered adipose tissue differentiation. We showed that some cultured skin fibroblasts issued from FPLD2 patients presented nuclei with abnormal shape with blebs and an abnormal repartition of lamin A/C and B [19]. Then, the presence of nuclear blebs has been reported in cells from all patients with mutations in LMNA and are now considered as a hallmark of laminopathies. Interestingly, we have recently shown that adipose tissue issued from hypertrophic neck fat presented a number of abnormalities indicating the presence of a mitochondrial dysfunction together with increased fibrosis. Adipocytes were not hypertrophied as expected but on the contrary reduced in size [20]. Therefore, *LMNA* mutations result in all fat depots in abnormal adipose tissue with defective differentiation.

In addition to the canonical *LMNA* mutations observed in patients with the typical form of FPLD2, several *LMNA* mutations have been reported in patients with atypical forms of lipodystrophies but severe insulin resistance. Since, in some of these patients, lipodystrophy is very slight, this new syndrome was denominated 'metabolic laminopathies' [8]. Some patients presented other signs of laminopathies with muscle and/or cardiac alterations. Their diagnosis is not obvious since these patients resemble those with the common metabolic syndrome. However, it is important to diagnose them in order to prevent early complications including cardiac rhythm or conduction disturbances that can be present in patients with typical FPLD2 [21]. In addition, since the disease is generally dominant, it

          Capeau·Magré·Caron-Debarle·Lagathu·Antoine·Béréziat·Lascols·Bastard·Vigouroux

is important to perform a familial screening in order to provide the patients with an adequate and early treatment. We propose to screen for mutations in *LMNA* in patients with lipodystrophy, even mild, if familial antecedents are present or in the case of associated muscular signs or cardiac disturbances. CT scan of the abdomen and/or thigh can help diagnosis so as the regional evaluation of body fat amount by DEXA.

The discovery in 2003 that a syndrome of premature aging called the Hutchinson-Gilford progeria resulted from a mutation in *LMNA* opened a new field of investigation on the role of lamin A/C [17]. Importantly the *LMNA* G608G mutation does not modify the postulated sequence of prelamin A but alters a splice site resulting in the deletion of 50 amino acids including the site of proteolysis by the protease ZMP-STE24. Therefore, this truncated prelamin A, called progerin, remains farnesylated and strongly anchored in the nuclear membrane. In addition to mutations in the lamin gene being responsible for progeria, other diseases with severe premature aging were discovered due either to mutations in *LMNA* or in the gene encoding ZMP-STE24 resulting in an increased level of farnesylated prelamin A. A number of studies tried to understand why the presence of progerin or farnesylated prelamin A is toxic for the cell. Some data outlined the important role of lamin A/C in the recruitment of DNA repair factors [22]. When altered, this results in genomic instability and p53 activation leading to cell senescence and accelerated aging. Very recent studies revealed that progerin could affect mesenchymal stem cells leading to altered differentiation of the cell lineage issued from these cells [23] including bone, muscle, adipose tissue, skin, all tissues affected in priority in laminopathies.

Interestingly, lipodystrophy, insulin resistance and early cardiovascular complications are features common both to premature aging syndromes and to typical or atypical lamin-linked lipodystrophies, suggesting some similar pathophysiological mechanisms. We thus searched for the presence of prelamin A in fibroblasts from patients with metabolic laminopathies and FPLD2. Farnesylated prelamin A was indeed present and *LMNA*-mutated cells presented features of early senescence [24].

Familial Partial Lipodystrophies Linked to Mutations in PPARG

The first patient described with a mutation in the gene encoding PPARγ was identified with a severe insulin resistance and hypertension but lipodystrophy, which was mild, was only diagnosed secondarily [25]. Since that, several patients with *PPARG* mutations were recognized, all characterized by mild forms of lipodystrophy, affecting the lower limbs and the buttocks but sparing the abdominal SAT, together with severe hypertension and metabolic abnormalities. About 15 different mutations have been identified, all heterozygous, leading to a dominant transmission of the disease [6].

The pathophysiological mechanisms involve alterations of the transcriptional activity of PPARγ, which is important for adipose tissue differentiation, but also plays a role in other tissues. The mutations identified could induce a dominant-negative effect but could also be deleterious due to haploinsufficiency [6]. The fact that patients with FPLD mutated on *PPARG* present a less severe lipodystrophy than those mutated on *LMNA* but more severe metabolic alterations suggests that PPARγ plays important roles outside adipose tissue.

Other Partial Inherited Lipodystrophies
A third gene encoding the protein kinase B, *AKT2*, has been involved in FPLD. This kinase, playing an important role as an intermediate in insulin signaling, was mutated in several members of a family with hyperinsulinemia and diabetes, indicating a dominant transmission of insulin resistance [26]. The proband presented with partial lipodystrophy [9]. However, this gene was not found mutated in other patients and two other missense mutations in *AKT2* do not clearly segregate with insulin resistance in the families and do not alter Akt2 kinase activity. In addition, heterozygous *CAV1* frameshift mutations have been reported in atypical forms of partial lipodystrophy. However, the phenotype is heterogeneous and the functional consequences of the different genetic alterations difficult to understand [3].

A number of patients with familial forms of partial lipodystrophies are at present undiagnosed. Recent studies from a large international collaboration indicate that some other genes could be involved in a few of them. Nevertheless, a number of cases remain unidentified at the genetic level. CIDEC was very recently recognized [41].

Acquired Lipodystrophies

The occurrence of lipodystrophy can be delayed in some genetic forms of lipodystrophy such as FPLD2, in which the phenotype appears after puberty. Otherwise, in some patients, lipodystrophy occurs during childhood and adulthood, without familial antecedents or mutations in genes known to be responsible for lipodystrophy, and in the context of an acute disease. At the clinical level, lipodystrophy can be complete or partial, very similar to that observed in genetic forms. Metabolic alterations are also very similar, arguing for the causative role of fat loss in metabolic disturbances.

Generalized Acquired Lipodystrophy
Also called the Lawrence syndrome, lipodystrophy occurs during childhood or adulthood sometimes preceded by an acute viral illness. The origin is unknown. However, in a number of patients, fat disappearance is preceded by a local inflammatory panniculitis. Also, signs of autoimmunity are sometimes present [27].

Therefore, in some patients, it is possible that fat is aggressed by an immune process leading to adipocyte destruction: when searched for, the presence of anti-adipocyte immune reactivity in patients' serum was found occasionally. Otherwise, the clinical and biological signs and the complications are very similar to those observed in BSCL [5, 6].

Acquired Partial Lipodystrophy
Among the different forms of partial acquired lipodystrophy, with no known etiology, one form can be identified since its phenotype is the reverse of that found in FPLD. Patients with the Barraquer-Simmons syndrome, more frequent in women, present a normal or decreased amount of fat in the upper part of the body (face, upper part of the trunk, arms) while fat in the lower part is in excess (buttocks, hips, legs). The etiology is unknown, even if autoimmunity has been reported in some cases. A membranoproliferative glomerulonephritis affects one third of the patients and more than half of them show signs of activation of the alternative complement pathway: low circulating levels of C3 and presence of C3 nephritic factor. Heterozygous alterations in *LMB2* encoding lamin B2 have been reported by the group of Hegele et al. [6] but were not confirmed by the other groups involved in the genetics of lipodystrophies.

Interestingly, while patients with loss of fat in the lower part of the body present severe metabolic alterations, patients with the reverse phenotype do not generally present such alterations, in agreement with the beneficial role of lower limbs fat at the metabolic level.

Lipodystrophies Linked to HIV Infection
Patients with HIV infection have benefitted from different classes of antiretroviral drugs since the late 1990s that has resulted in the control of the viral infection in most cases. In particular, the introduction in 1996 of the class of HIV protease inhibitors (PI) given in association with the class of nucleoside reverse transcriptase inhibitors (NRTI) led, in most patients, to an efficient control of the viral infection in the long term together with immune recovery of the number of CD4 T lymphocytes. However, at the time when PIs were introduced, a number of patients underwent lipodystrophy, with severe peripheral lipoatrophy and, in some patients, excessive visceral fat accumulation. These alterations in fat distribution were associated with metabolic disturbances, insulin resistance, diabetes, dyslipidemia [4–6, 28].

The prevalence of HIV-related lipodystrophies in the early 2000s was very high, affecting more than half of the patients and about 70–80% in some groups, such as the French ANRS APROCO cohort.

The pathophysiology of this lipodystrophy led to extensive clinical and fundamental studies [29]. A number of researches evaluated the ability of individual antiretroviral drugs to alter adipocyte functions in cultured cells. At first, the deleterious impact of first-generation PI such as nelfinavir and indinavir was clearly demonstrated, these drugs being able to inhibit differentiation, induce insulin resistance and increase the production by adipocytes of pro-inflammatory cytokines [30]. More recently, second-generation PIs were also evaluated for their ability to modify adipocyte phenotype: ritonavir and lopinavir exerted deleterious effects through the induction of an oxidative stress and the modification of adipokine secretion while atazanavir and amprenavir used alone were mainly devoid of an effect in that setting [31]. The ability of some PIs to induce IL-6 secretion through the activation of the NF-κB pathway was seen in human adipose tissue explants. Importantly, this effect was observed in explants from SAT but not VAT, in accordance with the clinical observation of preferential atrophy of SAT in HIV-infected patients [32]. A major pathway, which could also explain the effect of PIs, results from their ability to inhibit the enzyme ZMP-STE24, therefore leading to the accumulation of farnesylated prelamin A, to increased oxidative stress and to induction of an early cellular senescence [24]. This point is important to consider given the phenotype of premature aging which is frequently observed in HIV-infected patients.

The second class of antiretroviral drugs which is now suspected to play the leading role in lipoatrophy is the class of thymidine analogue NRTI and in particular the two first very active molecules, stavudine and zidovudine. These drugs were able to markedly alter adipocyte function in vitro through altered mitochondrial potential and increased oxidative stress [33] leading to decreased adiponectin secretion. They also increased MCP-1 and IL-6 production [31]. Thus, these two NRTIs but not the second-generation NRTIs were able to induce cellular premature senescence [33].

Clinical studies clearly revealed that the two thymidine NRTIs, mainly stavudine, were involved in priority in patients' lipoatrophy. PI could act in synergy with these NRTIs but could also impact on visceral fat and induce fat hypertrophy together with metabolic alterations, some PI being deleterious on lipids, with increased VLDL production by the liver, while others affect in priority insulin sensitivity and glucose metabolism. The demonstration of the toxic effects of antiretroviral drugs on adipose tissue was investigated in HIV-infected patients who were able to stop any antiretroviral treatment for at least 6 months. In this ANRS study, Lipostop, initial adipose tissue inflammation was markedly decreased, this improvement being related to the interruption of stavudine or zidovudine. However, both PI and thymidine analogues negatively impacted on different other adipocyte function [34].

More recently, a role for HIV infection has been also postulated, given that HIV can act on adipocytes, possibly by infecting them when adipocytes are in an inflammatory context [35] but also through the release by infected resident macrophages inside adipose tissue of viral proteins, which alter adipocyte phenotype

    Capeau · Magré · Caron-Debarle · Lagathu · Antoine · Béréziat · Lascols · Bastard · Vigouroux

[36]. In addition, when infected by HIV, macrophages shift their phenotype from a mainly anti-inflammatory M2 towards a M1 phenotype, resulting in the release of pro-inflammatory cytokines [37] and thereby in adipocyte insulin resistance and altered adipokine production. Therefore, taken as a whole, the severity of HIV-related lipodystrophy observed when patients were treated with first-generation antiretrovirals, could result from different factors, all aggressing adipocytes: the simultaneous use of drugs able to negatively impact on adipose tissue but also the long-term HIV infection, with probably, in most patients, the constitution of virus reservoirs in macrophages inside adipose tissue.

At present, the toxicity of the new drugs in the different antiretroviral classes is markedly decreased and the occurrence of the lipodystrophic phenotype reduced in HIV-infected patients. In most patients switched from first- towards second-generation antiretrovirals, lipodystrophy improved but the reversion is slow and sometimes incomplete. Plastic surgery could provide a valuable improvement at least for facial lipoatrophy, when severe.

However, HIV-infected patients, even well controlled with an undetectable viral load and a high number of CD4 lymphocytes, encounter the early occurrence of a number of comorbidities classically associated with aging: increased cardiovascular disease, hypertension, osteoporosis, neurocognitive decline, dyslipidemia, diabetes, renal and liver failure, sarcopenia and motor decline, frailty, non-acquired immunodeficiency syndrome defining malignancies. These alterations could reveal the occurrence in these patients of premature aging. Studies have to be performed to identify the reason for this process, which probably results from the chronic infection, leading to a long-term, low-grade inflammation, from immune senescence, leading to immune depletion, and from the adverse effect of some antiretroviral drugs. In addition, personal factors such as smoking, junk food, excessive alcohol intake and lack of exercise probably accentuate the infection-related alterations.

Lipodystrophies Linked to Excess Cortisol
It has long be recognized that patients with excess cortisol, either with a Cushing syndrome or disease or treated for a while with corticoids, present fat redistribution, with fat loss in the limbs and fat gain in the upper part of the trunk, including moon-like face, buffalo hump in the back and increased VAT. In addition, these subjects undergo bone loss, hypertension, hyperandrogenism and metabolic alterations with insulin resistance, linked to the effect of cortisol: altered glucose tolerance and even corticoid-induced diabetes, lipid alterations and increased cardiovascular risk. Therefore, this phenotype is clearly a cortisol-induced lipodystrophic syndrome.

The role of cortisol in adipose tissue is important to consider, since fat is able to convert inactive cortisone into active cortisol due to the presence of the enzyme 11β-hydroxysteroid dehydrogenase. It has been observed that abdominal SAT is able to secrete cortisol while the role of visceral fat in that setting has been recently questioned in men. Cortisol increases adipocyte size leading to insulin-resistant large adipocytes while it inhibits adipocyte proliferation. However, the mechanism by which cortisol induces lipodystrophy remains unclear.

And What about Aging?
During the process of physiological aging, fat depots present a physiologic redistribution towards central parts of the body, with loss of fat in the limbs and increased waist circumference, particularly seen in postmenopausal women. This redistribution is associated with age-related deterioration of insulin sensitivity, glucose and lipid parameters. In the context of increased body weight, this redistribution is more marked and the phenotype linked with central obesity has been individualized as the metabolic syndrome. Increased central fat, both subcutaneous and visceral, probably plays a leading role in the occurrence of the other alterations associated in the metabolic syndrome: decreased HDL, increased LDL cholesterol, TG, blood pressure, glycemia. Therefore, the metabolic syndrome could represent a mild form of acquired human lipodystrophy

Treatment of Lipodystrophies

The altered body fat repartition can benefit from plastic surgery. Patients with FPLD2 sometimes undergo successful removal of excess fat at the neck and face level. In patients with HIV-related facial lipoatrophy, plastic surgery is able to provide amelioration, even if often transitory: the Coleman technique consists in injection into the cheeks of autologous fat. Otherwise, correction with resorbable polyacrylamide gels or with non-resorbable fillers such as alkylamide allows partial correction. In some patients with severe hypertrophy of fat as a buffalo hump, fat removal can be proposed but with a risk of recurrence.

Some medications could possibly ameliorate lipoatrophy: a treatment with troglitazone, a first-generation thiazolidinedione (TZD), was initially shown to restore some fat in the limbs in patients with generalized lipodystrophy not related to HIV infection [38]. In patients with HIV-related lipodystrophy, TZD revealed poor efficacy on fat restoration. This was probably due to the presence of stavudine in the patients' treatment, which impeded fat restitution. When pioglitazone was given to patients not treated by stavudine, an improvement in peripheral fat was reported [39].

Treatment of metabolic alterations can benefit from diet recommendations that can ameliorate insulin sensitivity and hypertriglyceridemia. When diabetes is present, it is generally insulin-resistant and difficult to control. Insulin sensitizers are used at first, metformin and TZD. A treatment with TZD resulted in favorable effects on glucose control in several patients, even those with mutations in *PPARG*.

Very high doses of insulin are frequently required. In some cases, medium chain fatty acids supplementation can contribute to lower TG. Otherwise, hypolipidemic drugs are required.

Since these patients often present very low leptin levels, replacement of leptin with recombinant human leptin has been evaluated and in adults resulted in markedly improved metabolic values and regression of liver steatosis [40].

Conclusion

Lipodystrophies represent a heterogeneous group of severe diseases leading to early diabetic, cardiovascular and hepatic complications. Alterations in adipose tissue distribution could result from mutations in several genes: the presence of lipodystrophy outlines the importance of these genes in adipose tissue function. The role of lipid droplets as a new organelle playing a leading role in adipocyte functions is shown by the discovery that several genes mutated in lipodystrophies act at that level [3]. Active researches are looking for mutations in other candidate genes. Even if the pathophysiology of lipodystrophies remains largely unknown, it is obvious that all situations with fat loss, in particular in the lower body fat depots, are associated with severe metabolic disturbances and insulin resistance, while the only lipodystrophic syndrome with the reverse repartition of fat (the Barraquer-Simons syndrome) is generally not associated with metabolic alterations. This is reminiscent of the android obesity associated with abnormal metabolic parameters while the gynoid form is largely devoid of them.

Human partial lipodystrophies commonly associate loss of fat in some depots while others are increased. This points to the different physiology of the different fat depots, since the same genetic alteration or drug-induced toxicity results in opposite phenotypes depending on the fat localization.

The presence of mitochondrial dysfunction in lipodystrophies has been revealed in adipose tissue from patients with FPLD2 and other *LMNA* mutations but also in HIV-related lipodystrophies. Interestingly, some forms of lipomatosis result from mutations in mitochondrial DNA. Therefore, the relation between mitochondria and adipose tissue is probably important and complex and could result either in lipoatrophy but also in hypertrophied fat. Mitochondrial dysfunction has been also implied in muscular insulin resistance found during aging and in diabetic patients.

The specific role of lamin A/C in adipose tissue is important to consider. Accumulation of farnesylated prelamin A is involved in diseases associated with premature aging but also in *LMNA* and HIV-linked lipodystrophies, which also present signs of premature aging. During normal aging a fat redistribution is observed. Whether there is a link between lamin and normal aging remain to be demonstrated.

Therefore, studies on human lipodystrophies help to understand the complex physiology and pathology of fat. They point to new genes and new targets, which could lead to the discovery of new therapeutic clues in order to help treatment of patients with lipodystrophies but also, more generally, of patients with common forms of fat redistribution as observed in the metabolic syndrome and type 2 diabetes.

Acknowledgements

The authors were supported by grants from INSERM, ANRS, SIDACTION, ALFEDIAM, Fondation de France, and the European Eurolaminopathy Network.

References

1 Snijder MB, Dekker JM, Visser M, Bouter LM, Stehouwer CD, Yudkin JS, Heine RJ, Nijpels G, Seidell JC: Trunk fat and leg fat have independent and opposite associations with fasting and post-load glucose levels: the Hoorn study. Diabetes Care 2004;27:372–377.

2 Antuna-Puente B, Feve B, Fellahi S, Bastard JP: Adipokines: the missing link between insulin resistance and obesity. Diabetes Metab 2008;34:2–11.

3 Garg A, Agarwal AK: Lipodystrophies: disorders of adipose tissue biology. Biochim Biophys Acta 2009;1791:507–513.

4 Capeau J, Magré J, Lascols O, Caron M, Béréziat V, Vigouroux C, Bastard JP: Diseases of adipose tissue: genetic and acquired lipodystrophies. Biochem Soc Trans 2005;33:1073–1077.

5 Garg A: Acquired and inherited lipodystrophies. N Engl J Med 2004;350:1220–1234.

6 Hegele RA, Joy TR, Al-Attar SA, Rutt BK: Thematic review series: Adipocyte biology. Lipodystrophies: windows on adipose biology and metabolism. J Lipid Res 2007;48:1433–1444.

7 Semple RK, Cochran EK, Soos MA, Burling KA, Savage DB, Gorden P, O'Rahilly S: Plasma adiponectin as a marker of insulin receptor dysfunction: clinical utility in severe insulin resistance. Diabetes Care 2008;31:977–979.

8 Decaudain A, Vantyghem MC, Guerci B, Hecart AC, Auclair M, Reznik Y, Narbonne H, Ducluzeau PH, Donadille B, Lebbe C, Béréziat V, Capeau J, Lascols O, Vigouroux C: New metabolic phenotypes in laminopathies: LMNA mutations in patients with severe metabolic syndrome. J Clin Endocrinol Metab 2007;92:4835–4844.

9 Semple RK, Sleigh A, Murgatroyd PR, Adams CA, Bluck L, Jackson S, Vottero A, Kanabar D, Charlton-Menys V, Durrington P, Soos MA, Carpenter TA, Lomas DJ, Cochran EK, Gorden P, O'Rahilly S, Savage DB: Postreceptor insulin resistance contributes to human dyslipidemia and hepatic steatosis. J Clin Invest 2009;119:315–322.

10 Haque WA, Shimomura I, Matsuzawa Y, Garg A: Serum adiponectin and leptin levels in patients with lipodystrophies. J Clin Endocrinol Metab 2002;87:2395.

11 Savage DB, Petersen KF, Shulman GI: Disordered lipid metabolism and the pathogenesis of insulin resistance. Physiol Rev 2007;87:507–520.

12 Magré J, Delepine M, Khallouf E, Gedde-Dahl T Jr, Van Maldergem L, Sobel E, Papp J, Meier M, Megarbane A, Bachy A, Verloes A, d'Abronzo FH, Seemanova E, Assan R, Baudic N, Bourut C, Czernichow P, Huet F, Grigorescu F, de Kerdanet M, Lacombe D, Labrune P, Lanza M, Loret H, Matsuda F, Navarro J, Nivelon-Chevalier A, Polak M, Robert JJ, Tric P, Tubiana-Rufi N, Vigouroux C, Weissenbach J, Savasta S, Maassen JA, Trygstad O, Bogalho P, Freitas P, Medina JL, Bonnicci F, Joffe BI, Loyson G, Panz VR, Raal FJ, O'Rahilly S, Stephenson T, Kahn CR, Lathrop M, Capeau J: Identification of the gene altered in Berardinelli-Seip congenital lipodystrophy on chromosome 11q13. Nat Genet 2001;28:365–370.

13 Szymanski KM, Binns D, Bartz R, Grishin NV, Li WP, Agarwal AK, Garg A, Anderson RG, Goodman JM: The lipodystrophy protein seipin is found at endoplasmic reticulum lipid droplet junctions and is important for droplet morphology. Proc Natl Acad Sci USA 2007;104:20890–20895.

14 Boutet E, El Mourabit H, Prot M, Nemani M, Khallouf E, Colard O, Maurice M, Durand-Schneider AM, Chretien Y, Gres S, Wolf C, Saulnier-Blache JS, Capeau J, Magré J: Seipin deficiency alters fatty acid Δ^9-desaturation and lipid droplet formation in Berardinelli-Seip congenital lipodystrophy. Biochimie 2009;91:796–803.

15 Agarwal AK, Kazachkova I, Ten S, Garg A: Severe mandibuloacral dysplasia-associated lipodystrophy and progeria in a young girl with a novel homozygous Arg527Cys LMNA mutation. J Clin Endocrinol Metab 2008;93:4617–4623.

16 Kim CA, Delepine M, Boutet E, El Mourabit H, Le Lay S, Meier M, Nemani M, Bridel E, Leite CC, Bertola DR, Semple RK, O'Rahilly S, Dugail I, Capeau J, Lathrop M, Magré J: Association of a homozygous nonsense caveolin-1 mutation with Berardinelli-Seip congenital lipodystrophy. J Clin Endocrinol Metab 2008;93:1129–1134.

17 Worman HJ, Bonne G: 'Laminopathies': a wide spectrum of human diseases. Exp Cell Res 2007; 313:2121–2133.

18 Vigouroux C, Magré J, Vantyghem MC, Bourut C, Lascols O, Shackleton S, Lloyd DJ, Guerci B, Padova G, Valensi P, Grimaldi A, Piquemal R, Touraine P, Trembath RC, Capeau J: Lamin A/C gene: sex-determined expression of mutations in Dunnigan-type familial partial lipodystrophy and absence of coding mutations in congenital and acquired generalized lipoatrophy. Diabetes 2000; 49:1958–1962.

19 Vigouroux C, Auclair M, Dubosclard E, Pouchelet M, Capeau J, Courvalin JC, Buendia B: Nuclear envelope disorganization in fibroblasts from lipodystrophic patients with heterozygous R482Q/W mutations in the lamin A/C gene. J Cell Sci 2001; 114:4459–4468.

20 Béréziat V, Cervera P, Verpont M, Le Dour C, Antuna-Puente B, Dumont S, Somja-Azzi L, Vantyghem M, Capeau J, Vigouroux C: Adipose tissue of lipodystrophic patients carry mutations of the lamina A/C has fibrotic changes and mitochondrial alterations in the absence of inflammation. Diabetes Metab 2009;35 A27-A27.

21 Vantyghem MC, Pigny P, Maurage CA, Rouaix-Emery N, Stojkovic T, Cuisset JM, Millaire A, Lascols O, Vermersch P, Wemeau JL, Capeau J, Vigouroux C: Patients with familial partial lipodystrophy of the Dunnigan type due to a LMNA R482W mutation show muscular and cardiac abnormalities. J Clin Endocrinol Metab 2004;89: 5337–5346.

22 Dechat T, Pfleghaar K, Sengupta K, Shimi T, Shumaker DK, Solimando L, Goldman RD: Nuclear lamins: major factors in the structural organization and function of the nucleus and chromatin. Genes Dev 2008;22:832–853.

23 Scaffidi P, Misteli T: Lamin A-dependent misregulation of adult stem cells associated with accelerated ageing. Nat Cell Biol 2008;10:452–459.

24 Caron M, Auclair M, Donadille B, Béréziat V, Guerci B, Laville M, Narbonne H, Bodemer C, Lascols O, Capeau J, Vigouroux C: Human lipodystrophies linked to mutations in A-type lamins and to HIV protease inhibitor therapy are both associated with prelamin A accumulation, oxidative stress and premature cellular senescence. Cell Death Differ 2007;14:1759–1767.

25 Savage DB, Tan GD, Acerini CL, Jebb SA, Agostini M, Gurnell M, Williams RL, Umpleby AM, Thomas EL, Bell JD, Dixon AK, Dunne F, Boiani R, Cinti S, Vidal-Puig A, Karpe F, Chatterjee VK, O'Rahilly S: Human metabolic syndrome resulting from dominant-negative mutations in the nuclear receptor peroxisome proliferator-activated receptor-γ. Diabetes 2003;52:910–917.

26 George S, Rochford JJ, Wolfrum C, Gray SL, Schinner S, Wilson JC, Soos MA, Murgatroyd PR, Williams RM, Acerini CL, Dunger DB, Barford D, Umpleby AM, Wareham NJ, Davies HA, Schafer AJ, Stoffel M, O'Rahilly S, Barroso I: A family with severe insulin resistance and diabetes due to a mutation in AKT2. Science 2004;304:1325–1328.

27 Savage DB, Semple RK, Clatworthy MR, Lyons PA, Morgan BP, Cochran EK, Gorden P, Raymond-Barker P, Murgatroyd PR, Adams C, Scobie I, Mufti GJ, Alexander GJ, Thiru S, Murano I, Cinti S, Chaudhry AN, Smith KG, O'Rahilly S: Complement abnormalities in acquired lipodystrophy revisited. J Clin Endocrinol Metab 2009;94:10–16.

28 Saves M, Raffi F, Capeau J, Rozenbaum W, Ragnaud JM, Perronne C, Basdevant A, Leport C, Chene G: Factors related to lipodystrophy and metabolic alterations in patients with human immunodeficiency virus infection receiving highly active antiretroviral therapy. Clin Infect Dis 2002;34:1396–1405.

29 Gougeon ML, Penicaud L, Fromenty B, Leclercq P, Viard JP, Capeau J: Adipocytes targets and actors in the pathogenesis of HIV-associated lipodystrophy and metabolic alterations. Antivir Ther 2004;9:161–177.

30   Lagathu C, Kim M, Maachi M, Vigouroux C, Cervera P, Capeau J, Caron M, Bastard JP: HIV antiretroviral treatment alters adipokine expression and insulin sensitivity of adipose tissue in vitro and in vivo. Biochimie 2005;87:65–71.

31   Lagathu C, Eustace B, Prot M, Frantz D, Gu Y, Bastard JP, Maachi M, Azoulay S, Briggs M, Caron M, Capeau J: Some HIV antiretrovirals increase oxidative stress and alter chemokine, cytokine or adiponectin production in human adipocytes and macrophages. Antivir Ther 2007;12:489–500.

32   Vatier C, Leroyer S, Quette J, Brunel N, Capeau J, Antoine B: HIV protease inhibitors differently affect human subcutaneous and visceral fat: they induce IL-6 production and alter lipid storage capacity in subcutaneous but not visceral adipose tissue explants. Antiviral Ther 2008;13:A15-A16.

33   Caron M, Auclairt M, Vissian A, Vigouroux C, Capeau J: Contribution of mitochondrial dysfunction and oxidative stress to cellular premature senescence induced by antiretroviral thymidine analogues. Antivir Ther 2008;13:27–38.

34   Kim MJ, Leclercq P, Lanoy E, Cervera P, Antuna-Puente B, Maachi M, Dorofeev E, Slama L, Valantin MA, Costagliola D, Lombes A, Bastard JP, Capeau J: A 6-month interruption of antiretroviral therapy improves adipose tissue function in HIV-infected patients: the ANRS EP29 Lipostop Study. Antivir Ther 2007;12:1273–1283.

35   Maurin T, Saillan-Barreau C, Cousin B, Casteilla L, Doglio A, Penicaud L: Tumor necrosis factor-α stimulates HIV-1 production in primary culture of human adipocytes. Exp Cell Res 2005;304:544–551.

36   Giralt M, Domingo P, Villarroya F: HIV-1 infection and the PPARγ-dependent control of adipose tissue physiology. PPAR Res 2009;2009:607902.

37   Brown JN, Kohler JJ, Coberley CR, Sleasman JW, Goodenow MM: HIV-1 activates macrophages independent of Toll-like receptors. PLoS ONE 2008;3:e3664.

38   Arioglu E, Duncan-Morin J, Sebring N, Rother KI, Gottlieb N, Lieberman J, Herion D, Kleiner DE, Reynolds J, Premkumar A, Sumner AE, Hoofnagle J, Reitman ML, Taylor SI: Efficacy and safety of troglitazone in the treatment of lipodystrophy syndromes. Ann Intern Med 2000;133: 263–274.

39   Slama L, Lanoy E, Valantin MA, Bastard JP, Chermak A, Boutekatjirt A, William-Faltaos D, Billaud E, Molina JM, Capeau J, Costagliola D, Rozenbaum W: Effect of pioglitazone on HIV-1-related lipodystrophy: a randomized double-blind placebo-controlled trial (ANRS 113). Antivir Ther 2008;13:67–76.

40   Oral EA, Simha V, Ruiz E, Andewelt A, Premkumar A, Snell P, Wagner AJ, DePaoli AM, Reitman ML, Taylor SI, Gorden P, Garg A: Leptin-replacement therapy for lipodystrophy. N Engl J Med 2002;346:570–578.

41   Rubio-Cabezas O, Puri V, Murano I, Saudek V, Semple RK, Dash S, Hyden CS, Bottomley W, Vigouroux C, Magré J, Raymond-Barker P, Murgatroyd PR, Chawla A, Skepper JN, Chatterjee VK, Suliman S, Patch AM, Agarwal AK, Garg A, Brroso I, Cinti S, Czech MP, Argente J, O'Rahilly S, Savage DB, LD Screening Consortium: Partial lipodystrophy and insulin resistant diabetes in a patient with a homozygous nonsense mutation in CIDEC. EMBO Mol Med 2009;1:280–287.

J. Capeau
CDR Saint-Antoine, UMR S938, Faculty of Medicine Saint-Antoine
27, rue Chaligny, FR–75571 Paris Cedex 12 (France)
Tel. +33 1 40011332, Fax +33 1 40011432
E-Mail jacqueline.capeau@inserm.fr

Levy-Marchal C, Pénicaud L (eds): Adipose Tissue Development: From Animal Models to Clinical Conditions.
Endocr Dev. Basel, Karger, 2010, vol 19, pp 21–30

The Emergence of Adipocytes

Patrick Laharrague[a,b] · Louis Casteilla[b]

[a]Laboratoire d'Hématologie, CHU Toulouse, Hôpital Rangueil et [b]Métabolisme, Plasticité et Mitochondrie,
Université de Toulouse, Toulouse, France

Abstract

In mammals, the adipose organ is composed of white adipocytes (primary site in energy storage)
and of brown adipocytes (specialized in thermogenesis). Adipocytes arise from mesenchymal
stem cells (MSCs) by a sequential pathway of differentiation. MSCs develop either from ectoderm
or mesoderm and commit into different undifferentiated precursors, which upon the expression
of key transcription factors enter a differentiation program to acquire their specific functions.
When triggered by appropriate developmental cues, MSCs become committed to the adipocyte
lineage. White adipocytes differentiate from various types of vascular cell types, probably located
within the white adipose tissue itself. Brown adipocytes arise from myogenic precursors. The differentiation between white adipocyte and brown adipocyte lineages occurs in the earliest steps
of the fetal development, and both phenotypes are acquired independently. A better knowledge of these differentiation pathways allows new therapeutic strategies for reconstruction of
damaged conjunctive tissues and for the control or prevention of risks associated with obesity in
humans.

The Adipose Organ

Nearly all animal species have developed strategies to store excess energy in the
form of fat for later use. Worms store fat in intestinal epithelium and sharks in the
liver, both tissues of endodermal origin. But in most species, fat storage occurs in
an a priori mesodermal tissue, the white adipose tissue (WAT). The distribution
of white tissue varies between species: for invertebrates, amphibians, and numerous reptiles, the largest stores are intra-abdominal; most mammals and birds have
both intra-abdominal and subcutaneous fat pads [1]. This distribution does not
simply reflect an evolutionary adaptation for heat insulation since similar adipose
location is observed in artic and tropical mammals of similar weight [2].

In mammals, energy storage changes with age. The continuous and active
transfer of nutrients through the placenta fulfils the energetic demands, growth

and fat storage of the fetus. The latter is highly variable among mammals, lipid storage during fetal life being an exception rather a rule, and in most species body fat content at birth is very low [3]. In the human newborn, fat represents 16% of body weight. Body fat accretion occurs essentially during the last trimester of intrauterine life.

In mammals, the adipose organ is composed not only of white adipocytes (primary site in energy storage and mobilization in the form of triglyceride), but also of brown adipocytes (specialized in basal and inducible energy dissipation as thermogenesis). This thermogenesis, which is involved in diet induced and non-shivering thermogenesis occurs through expression of uncoupling protein-1 (UCP-1), an inner mitochondrial membrane protein that allows dissipation of the proton electrochemical gradient generated by respiration in the form of heat [4].

Although some warm-bodied fishes maintain a brain temperature higher than ambient temperature due to the presence of brown adipocyte-like cells around the brain, brown adipose tissue (BAT) itself develops relatively late in the course of evolution, in parallel with the development of homeothermy and the capacity for non-shivering thermoregulation [1]. No BAT is observed in amphibians and reptiles. Birds and pigs, though homeotherms, are devoid of BAT, and depend on muscular UCP for thermoregulation.

In rodents, BAT is most abundant in the neonatal period and is most concentrated in the interscapular region. Brown adipocytes can also be found in other areas, including typical WAT pads, following cold exposure. In human fetuses and newborns, BAT is found essentially in axillary, cervical, perirenal, and periadrenal regions [5]. In large mammals, BAT decreases shortly after birth [6]. It has classically been considered insignificant in human adults (except in patients with pheochromocytoma and in outdoor workers subjected to prolonged cold exposure). However, recent morphological and scanning studies have detected metabolically active brown fat in the cervical, supraclavicular, axillary, and paravertebral regions of normal individuals [7–9]. In addition, UCP-1 mRNA can be detected in human WAT, particularly after treatment by the antidiabetic drug thiazolidinedione, suggesting some admixture of BAT in WAT depots, and confirming our previous observations in rodents [10].

In adult human, WAT is dispersed throughout the body with major intra-abdominal (around the omentum, intestines, and perirenal areas) and subcutaneous depots (buttocks, thighs, and abdomen). In addition, WAT is found in many other areas: retro-orbital space, face, extremities, bone marrow. Some adipose tissue is responsive to sex hormones (such as in the breasts and thighs), whereas other depots (neck and upper back) are more responsive to glucocorticoids [1]. Differences in the metabolic properties and patterns of gene expression within different fat depots have also been described. Fat distribution influences risks associated with obesity in humans. Obese individuals with pear-shaped obesity (increased subcutaneous fat) are at low risk of diabetes and metabolic syndrome; individuals with apple-shaped obesity (increased visceral fat) are at high risk for metabolic complications of obesity. Finally, there is a genetic control of body fat distribution, evident in Hottentot/Khoisan women with steatopygia, but also observed in individuals with inheritable forms of partial lipodystrophy [11]. Using transplantation of internal or subcutaneous fat, Gesta et al. [12] nicely demonstrated that subcutaneous fat is intrinsically different from visceral fat and protects from metabolic disorders. These effects could be mediated by secreted substances that can act systemically to improve glucose metabolism.

The factors determining fat mass in adult humans are not fully understood, but increased lipid storage in already developed adipocytes is thought to be most important. The number of adipocytes stays constant in adulthood in lean and obese individuals, even after marked weight loss, suggesting that the number of fat cells is set during childhood. However, the fat organ is not fixed: approximately 10% of fat cells are renewed annually at all adult ages and levels of body mass index, indicating a high turnover of adipocytes [13]. This property is also indirectly revealed by the great sensibility of adipose tissues to irradiation [14]. Fat distribution changes with age, even in thin individuals with steady body weight: decrease in retro-orbital and subcutaneous depots, increase in intra-abdominal depots.

Developmental Origin of Fat

Like muscle and bone, adipose tissue is generally regarded as having a mesodermal origin, even if precise lineage tracing studies have not been performed. The formation of the mesoderm begins with the migration of a layer of cells between the primitive endoderm and ectoderm. This cell layer spreads along the antero-posterior and dorsoventral axes of the embryo giving rise to the axial, intermediate, lateral plate, and paraxial mesoderm. The latter, after its segmentation into somites, gives rise to the axial skeleton and muscles of the trunk. The lateral plate mesoderm generates the skeleton and muscles of the limbs. However, the bones and muscles of the skull and face are of ectodermal origin, specifically the neural crest [15]. Yet it was recently reported that neural crest stem cells, upon stimulation in vitro with defined factors, are able to differentiate into adipocytes, and in vivo using *Cre*-mediated recombination in transgenic mice, that a subset of adipocytes originates from the neural crest during development [16]. Each of these regions could give rise to local adipose tissue. This dual origin (mesoderm and ectoderm) is consistent with the differential expression of phenotypes and gene patterns between different fat pads and with the occurrence of various forms of partial dystrophy.

The mesodermal origin of BATs is clearly demonstrated. When mesoderm from a 9-day-old rat embryo is engrafted below the kidney capsule of an adult rat, it develops only into BAT [17]. Additionally, Atit et al. [18], using lineage-tracing techniques, demonstrated that some interscapular BAT bundles originate from the paraxial mesoderm.

Origin of Adipocytes

Very few data are available concerning the early development of adipose tissue in humans. Precise lineage tracing studies have not been performed, except for BAT. Most of our knowledge concerning WAT comes from histological observations.

In rodents, WAT develops mainly after birth. In humans, WAT formation begins during the second trimester of gestation. Wassermann [19] in 1965 studied for the first time the development of WAT in comparison with other organs. Through a careful histological study he demonstrated that adipose depots develop from primitive organs.. Within these primitive organs, clusters of adipocytes emerge from a bulk of mesenchymal cells related to the development of the vascular network, giving rise to fatty lobules. The most differentiated cells are far distant from capillaries. Vascularization therefore plays a major role in the development of adipose tissues. Angiogenesis and adipogenesis appear coordinated in time and space. We further discuss the particular relationship between adipose cells and endothelial cells.

Adipocytes arise from mesenchymal stem cells (MSCs) by a sequential pathway of differentiation. MSCs are commonly defined by plastic adherent growth, by a panel of surface markers (positive for CD73+ and CD105+, negative for hematopoietic makers), and by their in vitro capacity to enter the osteogenic, adipogenic, and chondrogenic lineages.

MSCs develop from the mesoderm and then commit into different lineages influenced by a number of factors. Bone morphogenetic proteins (BMPs) through their intracellular mediators (Smad proteins) can trigger MSCs to differentiate into osteogenic or adipogenic lineages, while preventing commitment into the myogenic lineage. Intracellular proteins, such as TAZ and Shn, modify the action of BMPs in the determination of the osteogenic or adipogenic lineages. In addition, BMP scan be modulated by Noggin, Nodal and glypican 3. Wnt and Hedgehog proteins are important to for myogenic and osteogenic commitment and prevent adipogenic differentiation. Comparative transcriptomic analyses reveal gene repression as a predominant early mechanism before final cell commitment [20]. Once committed, MSCs give rise to undifferentiated precursors (osteoblasts, adipoblasts/preadipocytes, myoblasts), which upon the expression of key transcription factors enter a differentiation program to acquire their specific functions (fig. 1).

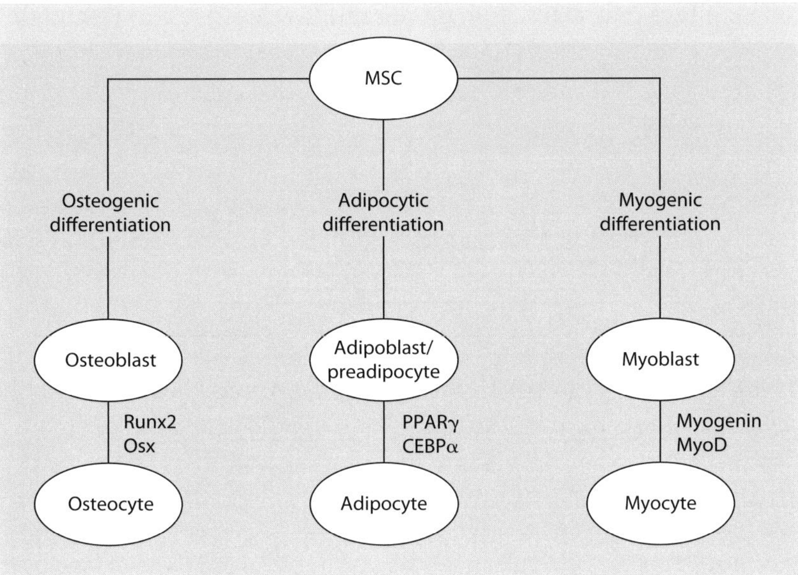

Fig. 1. Development pathways of mesenchymal lineages. MSCs were initially identified in bone marrow and have been used to model differentiating mesoderm. MSCs commit into different lineages, then give rise to undifferentiated precursors, which in turn acquire their definitive phenotype upon the expression of specific transcription factors.

The human MSC secretome at early steps of adipocyte differentiation was recently characterized [21]. Eight clusters have been identified, including proteases, protease inhibitors, ECM components, anti-inflammatory-antioxidant proteins, metabolic enzymes, cytoskeletal components, and heat-shock/protein folding proteins.

When triggered by appropriate developmental cues, MSCs become committed to the adipocyte lineage. A specificity of this developmental program is the pivotal role played by nutriments, either glucose or lipids, not only as metabolite substrates but also as true adipogenic signals. However, adipogenic differentiation pathway appears as a default pathway because always inhibited by master genes driving other differentiation pathways to give rise mesodermal lineages [22].

We established that many of adipose precursors express the surface marker CD34, a protein also present at the surface of immature cells and endothelial cells [23]. Since we have also provided evidence of a true angiogenic potential in vitro and in vivo, we hypothesized that these precursors could committed to endothelial lineage according appropriate conditions. A similar conclusion was drawn by Bouloumié et al. [24] from in vivo studies. More recently, employing a non-invasive assay for following fat mass reconstitution in vivo, Friedman et al. [25] identified a subpopulation of early adipocyte progenitor cells (Lin–, CD29+, CD34+, Sca-1+, CD24+) resident in adult WAT. When injected into the residual fat pads of lipodystrophic mice, these cells reconstitute a normal WAT depot and rescue the diabetic phenotype that develops in these animals. Using genetically marked mice, Graff et al. [26] found that most adipocytes descend from a pool of proliferating progenitors that are already committed, either prenatally or early in postnatal life. These progenitors reside in the mural cell compartment of the adipose vasculature, but not in the vasculature of other tissues. These data could be related to the observations made by an international team coordinated by Péault et al. [27] in Pittsburgh documenting a subset of human perivascular cells that express both pericyte and MSC markers in situ (CD146, NG2, PDGF-Rβ). Therefore, white adipocytes could differentiate from various types of vascular cell types, probably located within the WAT itself. However, it seems that the developmental origin of white preadipocytes is different according to the location [28].

What about brown adipocytes? In vitro, whereas some members of the family of BMPs support white adipocyte differentiation (BMP2 and BMP4, particularly), BMP7 promotes differentiation of brown preadipocytes. BMP7 activates a full program of brown adipogenesis including specific regulators such as *Pgc-1α* and *Ucp1*. Moreover, BMP7 triggers commitment of mesenchymal progenitor cells to a brown adipocyte lineage, and implantation of these cells into nude mice results in development of adipose tissue containing mostly brown adipocytes. Conversely, *BMP7* knockout embryos show a marked paucity of BAT and almost complete absence of UCP1 [29].

In vivo, two recent studies provide evidence for a close relationship between brown fat and skeletal muscle in development. Sorting cells from various tissues and differentiating them in an adipogenic medium, Giacobino et al. [30] reveal that a stationary population of skeletal muscle cells expressing the CD34 surface protein can differentiate into brown adipocytes with a high level of UCP1 expression and uncoupled respiration. Using fate mapping in mouse, Spiegelman et al. [31] show that brown – but not white – fat cells and skeletal muscle develop from a common progenitor that expresses the transcription factor myf5. The transcriptional regulators PRDM16 specify the brown fat lineage from the myf5-expressingg progenitors through mechanisms that involve activation of PPARγ and suppression of myogenic factors. Loss of PRDM16 from brown fat precursors disrupts their differentiation into BAT and enhances expression of muscle genes; in contrast, ectopic expression of PRDM16 in myoblasts induces brown fat adipogenesis. However, the brown fat cells that emerge in WAT in response to chronic cold exposure are not derived from myf5-expressing progenitors. If confirmed, these observations reveal the existence of two types of brown adipocytes, arising from distinct developmental pathways.

On the basis of the observations of Spiegelman et al. [31], it is likely that brown adipocytes arise from myogenic precursors in response to unknown effectors that enhance PRDM16 expression. In different strains of mice, the abundance of brown

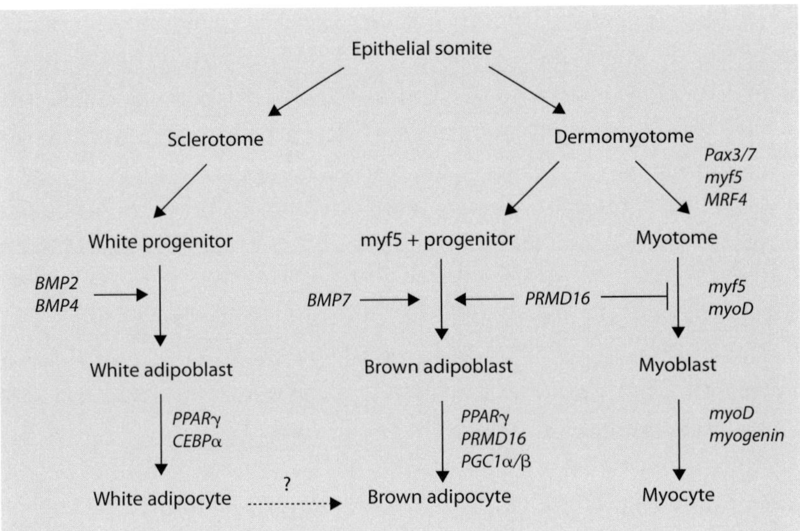

Fig. 2. Development pathways for WAT, BAT, and muscle. Recent studies identify a population of progenitor cells common to brown fat and skeletal muscle but not white fat. These progenitors express the transcription factor myf5 and arise within the developing dermomyotome. The transcription factor PRDM16 activates expression of the nuclear hormone receptor PPARγ. Together, PRDM16 and PPARγ suppress myogenesis and initiate development of brown adipocytes. The mechanism of early development of white adipocyte is unknown, although progenitors may arise from the developing sclerotome. The emergence of brown adipocytes in WAT is not presently known.

adipocytes correlates with energy expenditure and resistance to obesity [32]. It is quite possible that these brown adipocytes in muscle develop from myf5-expressing satellite cells (myogenic precursors) that principally function to repair damaged muscle tissue, but could also enhance the oxidative capacity of skeletal muscle [33].

The differentiation between white adipocyte and brown adipocyte lineages could occur in the earliest steps of the fetal development (fig. 2): emerging from epithelial somites the white progenitors could derive from the sclerotome, since brown progenitors/myoblasts could derive from the dermomyotome [33]. Some years ago, using transgenic models, we already demonstrated that white and brown phenotypes are acquired independently in the course of normal development [34].

Conclusion

Like other cellular types, such as fibroblasts or macrophages, with various phenotypes but sharing a common function, adipocytes constitute a heterogeneous and plastic population. Their common characteristic is participation to the energetic metabolism. For many years a distinction was made between white adipocytes (primary site of energy storage as triglycerides) and brown adipocytes (specialized for basal and inducible energy expenditure). Then between subcutaneous and intra-abdominal fat depots, with significant differences in cellular composition, secretion of adipokines, gene expression, and physiological properties [35]. Likewise, increased visceral fat appears associated to a high risk of insulin resistance and cardiovascular diseases, since individuals with increased subcutaneous fat are at low risk of metabolic complications.

In the same time, it was demonstrated that, in addition to adipocytes and preadipocytes, adipose tissues contain progenitor cells with in vitro and in vivo potential to differentiate into conjunctive cells [36], then representing a reservoir of cells for reconstructive therapy [37, 38].

We previously indicated that brown and white adipocytes have a distinct origin during development, and that white adipocytes derived either from ectoderm or mesoderm. Therefore, one important but still unsolved question relates to defining exactly the link between origin and physiological or physiopathological properties of adipocytes.

The question concerning the origin of adipocytes is not only of a cognitive interest. A better knowledge of the distinct differentiation pathways (pericytes vs. mesenchymal stem cells, osteoblasts vs. white preadipocytes, and myoblasts vs. brown preadipocytes) allows new therapeutic strategies. Not only with the isolation and expansion of cells for reconstruction of damaged conjunctive tissues, but also with the detection of new therapeutic targets for the control or prevention of risks associated with obesity in humans.

References

1   Gesta S, Tseng YH: Developmental origin of fat: tracking obesity to its source. Cell 2007;131:242.
2   Pond CM: An evolutionary and functional view of mammalian adipose tissue. Proc Nutr Soc 1992;51:367.
3   Herrera E, Amusquivar A: Lipid metabolism in the fetus and the newborn. Diabetes Metab Res Rev 2000;16:202.
4   Himms-Hagen J: Brown adipose tissue thermogenesis: interdisciplinary studies. FASEB J 1990; 4:2890.
5   Cannon B, Nedergaard J: Brown adipose tissue: function and physiological significance. Physiol Rev 2004;84:225.
6   Casteilla L, Champigny O, Bouillaud F, Robelin J, Ricquier D: Sequential changes in the expression of mitochondrial protein mRNA during the development of brown adipose tissue in bovine and ovine species. Sudden occurrence of uncoupling protein mRNA during embryogenesis and its disappearance after birth. Biochem J 1989; 257:665.
7   Nedergaard J, Bengtsson T, Cannon B: Unexpected evidence for active brown adipose tissue in adult humans. Am J Physiol 2007;293:E444.
8   Cypess AM, Lehman S, Williams G, Tal I, Rodman D, Goldfine AB, Kuo FC, Palmer EL, Tseng YH, Doria A, Kolodny GM, Kahn CR: Identification and importance of brown adipose tissue in adult humans. N Engl J Med 2009;360:1509.

9 Van Marken Lichtenbelt WD, Vanhommerig JW, Smulders NM, Drossaerts JM, Kemerink GJ, Bouvy ND, Schrauwen P, Teule GJ: Cold-activated brown adipose tissue in healthy men. N Engl J Med 2009;360:1500.

10 Cousin B, Cinti S, Morroni M, Raimbault S, Ricquier D, Pénicaud L, Casteilla L: Occurrence of brown adipocytes in rat white adipose tissue: molecular and morphological characterization. J Cell Sci 1992;103:931.

11 Agarwal AK, Garg A: Genetic disorders of adipose tissue development, differentiation, and death. Annu Rev Genomics Hum Genet 2006;7:175.

12 Tran TT, Yamamoto Y, Gesta S, Kahn CR: Beneficial effects of subcutaneous fat transplantation on metabolism. Cell Metab 2008;7:410.

13 Spalding KL, Arner E, Westermark PO, Bernard S, Buchholz BA, Bergmann O, Blomqvist L, Hoffstedt J, Näslund E, Britton T, Concha H, Hassan M, Rydén M, Frisén J, Arner P: Dynamics of fat cell turnover in humans. Nature 2008;453:783.

14 Poglio S, Galvani S, Bour S, André M, Prunet-Marcassus B, Pénicaud L, Casteilla L, Cousin B: Adipose tissue sensitivity to radiation exposure. Am J Pathol 2009;174:44.

15 Bronner-Fraser M: Neural crest cell formation and migration in the developing embryo. FASEB J 1994;8:699.

16 Billon N, Iannarelli P, Monteiro MC, Glavieux-Pardanaud C, Richardson WD, Kessaris N, Dani C, Dupin E: The generation of adipocytes by the neural crest. Development 2007;134:2283.

17 Loncar D: Brown adipose tissue as a derivative of mesoderm grafted below the kidney capsule. J Dev Biol 1992;36:265.

18 Atit R, Sgaler SK, Mohamed OA, Taketo MM, Dufort D, Joyner AL, Niswander L, Conlon RA: Beta-catenin activation is necessary and sufficient to specify the dorsal dermal fate in the mouse. Dev Biol 2008;296:164.

19 Wassermann P: The Development of Adipose Tissue. Baltimore, Waverly Press, 1965.

20 Scheideler M, Elabd C, Zaragosi LE, Chiellini C, Hackl H, Sanchez-Cabo F, Yadav S, Duszka K, Friedl G, Papak C, Prokesch A, Windhager R, Ailhaud G, Dani C, Amri EZ, Trajanoski Z: Comparative transcriptomics of human multipotent stem cells during adipogenesis and osteoblastogenesis. BMC Genomics 2008;9:340.

21 Chiellini C, Cochet O, Negroni L, Samson M, Poggi M, Ailhaud G, Alessi MC, Dani C, Amri EZ: Characterization of human mesenchymal stem cell secretome at early steps of adipocyte and osteoblast differentiation. BMC Mol Biol 2008;9:26.

22 Baulande S, Feve B: Identification of new genes associated with adipogenesis. Med Sci (Paris) 2003;19:151.

23 Planat-Benard V, Silvestre JS, Cousin B, Andre M, Nibbelink M, Tamarat R, Clergue M, Manneville C, Saillan-Barreau C, Duriez M, Tedgui A, Levy B, Penicaud L, Casteilla L: Plasticity of human adipose lineage cells toward endothelial cells: physiological and therapeutic perspectives. Circulation 2004;109:656.

24 Sengenès C, Miranville A, Maumus M, de Barros S, Busse R, Bouloumié A: Chemotaxis and differentiation of human adipose tissue CD34+/CD31– progenitor cells: role of stromal derived factor-1 released by adipose tissue capillary endothelial cells. Stem Cells 2007;25:2269.

25 Rodeheffer MS, Birsoy K, Friedman JM: Identification of white adipocyte progenitor cells in vivo. Cell 2008;135:240.

26 Tang W, Zeve D, Suh JM, Bosnakovski D, Kyba M, Hammer RE, Tallquist MD, Graff JM: White fat progenitor cells reside in the adipose vasculature. Science 2008;322:583.

27 Crisan M, Yap S, Casteilla L, Chen CW, Corselli M, Park TS, Andriolo G, Sun B, Zheng B, Zhang L, Norotte C, Teng PN, Traas J, Schugar R, Deasy BM, Badylak S, Buhring HJ, Giacobino JP, Lazzari L, Huard J, Péault B: A perivascular origin for mesenchymal stem cells in multiple human organs. Stem Cells 2008;26:2425.

28 Gesta S, Blüher M, Yamamoto Y, Norris AW, Berndt J, Kralisch S, Boucher J, Lewis C, Kahn CR: Evidence for a role of developmental genes in the origin of obesity and body fat distribution. Proc Natl Acad Sci USA 2006;103:6676–6681.

29 Tseng Y-H, Kokkotou E, Schulz TJ, Huang TL, Winnay JN, Taniguchi CM, Tran TT, Suzuki R, Espinoza DO, Yamamoto Y, Ahrens MJ, Dudley AT, Norris AW, Kulkarni RN, Kahn CR: New role of bone morphogenetic protein 7 in brown adipogenesis and energy expenditure. Nature 2008;454:1000.

30 Crisan M, Casteilla L, Lehr L, Carmona M, Paolini-Giacobino A, Yap S, Sun B, Léger B, Logar A, Pénicaud L, Schrauwen P, Cameron-Smith D, Paul RA, Péault B, Giacobino J-P: A reservoir of brown adipocyte progenitors in human skeletal muscle. Stem Cells 2008;3:1.

31 Seale P, Bjork B, Yang W, Kajimura S, Chin S, Kuang S, Scimè A, Devarakonda S, Conroe HM, Erdjument-Bromage H, Tempst P, Rudnicki MA, Beier DR, Spiegelman BM: PRDM16 controls a brown fat/skeletal muscle switch. Nature 2008; 454:961.

32 Almind K, Manieri M, Sivitz WI, Cinti S, Kahn CR: Ectopic brown adipose tissue in muscle provides a mechanism for differences in risk of metabolic syndrome in mice. Proc Natl Acad Sci USA 2007;104:2366.

33 Farmer RS: Brown fat and skeletal muscle: unlikely cousins? Cell 2008;134:726.

34 Moulin K, Truel N, André M, Arnauld E, Nibbelink M, Cousin B, Dani C, Pénicaud L, Casteilla L: Emergence during development of the white-adipocyte cell phenotype is independent of the brown-adipocyte cell phenotype. Biochem J 2001; 356:659.

35 Sethi JK, Vidal-Puig AJ: Adipocyte biology: adipose tissue function and plasticity orchestrate nutritional adaptation. J Lipid Res 2007;48:1253.

36 Fraser JK, Wulur I, Alfonso Z, Zhu M: Wheeler ES: Differences in stem and progenitor cell yield in different subcutaneous adipose tissue depots. Cytotherapy 2007;9:459.

37 Gimble JM, Katz AJ, Bunnell BA: Adipose-derived stem cells for regenerative medicine Circ Res 2007;100:1249.

38 Casteilla L, Dani C: Adipose tissue-derived cells: from physiology to regenerative medicine. Diabetes Metab 2006;32:393.

Patrick Laharrague
CNRS, UMR 5241, Université de Toulouse, BP 84225
FR–31432 Toulouse Cedex 4 (France)
Tel. +33 5 62 17 09 04, Fax +33 5 62 17 09 05
E-Mail laharrague.p@chu-toulouse.fr

Levy-Marchal C, Pénicaud L (eds): Adipose Tissue Development: From Animal Models to Clinical Conditions.
Endocr Dev. Basel, Karger, 2010, vol 19, pp 31–44

Adipose Tissue and the Reproductive Axis: Biological Aspects

G.J. Hausman · C.R. Barb

USDA/ARS, Richard B. Russell Agriculture Research Center, Athens, Ga., USA

Abstract
The discovery of leptin has clearly demonstrated a relationship between body fat and the neuroendocrine axis since leptin influences appetite and the reproductive axis. Since adipose tissue is a primary source of leptin, adipose tissue is no longer considered as simply a depot to store fat. Recent findings demonstrate that numerous other genes, i.e. neuropeptides, interleukins and other cytokines and biologically active substances such as leptin and insulin-like growth factors I and II, are also produced by adipose tissue, which could influence appetite and the reproductive axis. Targets of leptin in the hypothalamus include neuropeptide Y, proopiomelanocortin and kisspeptin. Transsynaptic connection of hypothalamic neurons to porcine adipose tissue may result in a direct influence of the hypothalamus on adipose tissue function. Nutritional signals such as leptin are detected by the central nervous system and translated by the neuroendocrine system into signals which ultimately regulates luteinizing hormone secretion. Furthermore, leptin directly affects gonadotropin-releasing hormone release from the hypothalamus, luteinizing hormone from the pituitary gland and ovarian follicular steroidogenesis. Although leptin is identified as a putative signal that links metabolic status and neuroendocrine control of reproduction, other adipocyte protein products may play key roles in regulating the reproductive axis in the pig.

Maintenance of an adequate energy supply is important for support of the reproductive process. The brain receives metabolic signals which relay either a negative or positive energy balance. One of the most widely studied metabolic signals is leptin. Leptin not only affects feed intake, but also the neuroendocrine axis, metabolism and immunological processes [1–3]. Leptin was first identified as the gene product found deficient in the obese ob/ob mouse [4]. The hypothalamus appears to be the primary site of action, since leptin receptors are located within hypothalamic areas associated with control of appetite, reproduction and growth [5, 6]. The discovery of leptin has improved our understanding of the relationship between adipose tissue and energy homeostasis [3, 7]. Increased leptin production by adipose tissue and rising levels of triglyceride stores in adipose tissue could serve as a signal to the brain, to decrease food intake and increase energy expenditure and resistance to obesity [3]. Moreover, when energy intake and output are equal, leptin reflects the amount of stored triglycerides in adipose tissue. Thus, leptin may serve as a circulating signal of nutritional status or lipostat, first proposed by Kennedy [8] in 1953. Furthermore, leptin may act as an important regulator of appetite, energy metabolism, body composition and reproduction. The intent of this review is to examine the biological role of leptin and also to examine the emerging role of other adipokines in the regulation of energy homeostasis and reproduction (table 1).

Adipose Tissue as an Endocrine Organ

Identification of Other Putative Adipose-Derived Factors or 'Adipokines'
Are there other 'leptins' or adipose-derived factors that play a role in reproduction? Adipose tissue microarrays from growing pigs (90, 150, and 210 days of age; 5 per age group) were examined for potential new adipokines with total RNA collected at the time of slaughter from two adipose depots, namely the outer subcutaneous adipose tissue (OSQ) and middle subcutaneous adipose tissue (MSQ). Gene percentages calculated from growing pig adipose tissue microarrays indicated significant main effects of interleukins (ILs), interferons (IFNs) and transforming growth factor (TGF) family members (p < 0.0001) from the MSQ depot and for ILs (p < 0.001) and TGF genes (p < 0.0001) from the OSQ depot [9]. Furthermore, distinct patterns of relative gene expression are evident within apolipoproteins, ILs, IFNs and TGF family members in adipose tissue from growing pigs [9]. Patterns of gene expression within apolipoproteins, ILs, IFNs, and TGF family members also distinguished OSQ and MSQ depots in growing pigs. The expression of several major cytokine and apolipoprotein genes including small inducible cytokine A5 (RANTES), IL-1B, IL-1A, IL-12A, IL-1 receptor antagonist and apolipoproteins A_1 and E are detected in pig adipose tissue with microarray and RT-PCR assays [9]. These studies demonstrate that expression of major cytokine and apolipoprotein genes in pig adipose tissue are not influenced by age or attainment of puberty in growing pigs, but may be influenced by location or depot.

Distinct patterns of relative gene expression are evident within neuropeptide Y (NPY) receptor (NPYR) and IGF-binding protein (IGFBP) family members in adipose tissue from growing pigs [10]. Relative gene expression levels of NPY2R, NPY4R and angiopoietin 2 (ANG-2) distinguished OSQ and MSQ depots in growing pigs [10]. The expressions of IGFBP-7, IGFBP-5, NPY1R, NPY2R, NPY, connective tissue growth factor (CTGF), brain-derived neurotrophic factor (BDNF)

Table 1. Principal adipokines and their known or potential involvement in reproduction and/or metabolism

Adipose secreted factor	Relationship to adiposity	Known or predicted effects on reproduction or metabolism	References
Leptin	Increases with fat accumulation	Regulation of LH secretion and follicle steroidogenesis	2, 13, 23, 33, 35, 41–43
Adiponectin	Decreases with adiposity	Pleiotropic effects on ovulation, steroidogenesis and placental function	46, 49
Nesfatin	Unknown	Suppress appetite, and stimulated LH secretion	50, 60
Visfatin	Increased with visceral adiposity in rats No relationship to adiposity in pig	Increased insulin sensitivity	61, 62
TNFα	Increases with obesity	Role in corpus luteum function, preovulatory follicle	63–65
IL-6, cytokines	Increases with obesity	Involved with inflammatory processes, ovulation, steroidogenesis, apoptosis	66–68
CNTF	Unknown	Inhibits appetite, stimulated LH secretion	69–71
Angiopoietin-like protein-4	Correlation with adiposity	Regulates glucose homeostasis, lipid metabolism and insulin sensitivity	72, 73
Resistin	Increases with adiposity	Reduced insulin sensitivity	74, 75
Bone morphogenetic protein-15	No correlation with adiposity in the growing pig	Potent stimulator of granulosa cell proliferation and selective modulator of FSH action	9, 76–78

and ciliary neurotrophic factor (CNTF) genes are detected in pig adipose tissue with microarray and RT-PCR assays. Furthermore, adipose tissue CTGF gene expression is upregulated while NPY and NPY2R gene expression were significantly downregulated by age [10]. These studies demonstrate that expression of neuropeptides and neurotrophic factors in pig adipose tissue may be involved in puberty through the regulation of leptin secretion. Microarray and RT-PCR studies are collaborated by proteomic studies of adipose tissue S-V cell cultures which identified cytokines such as IL-6, IL-4, IL-1A, IL-8, brain morphogenetic protein (BMP)-4, IGFBP-3, RANTES, tumor necrosis factor (TNF)-α and neurotrophic agents, BDNF, CNTF and apolipoproteins at the protein level [11].

Gene Expression in Subcutaneous Adipose Tissue from Growing Pigs and Neonatal and Fetal Pigs

Analysis of MSQ depot IL percentages from growing pigs showed that IL-15 and IL-6 percentages are significantly different from each other; IL-15 is different from all other ILs and IL-6 different from nearly all other IL percentages [9]. Analysis of neonatal adipose tissue IL percentages showed that the IL-15 percentage is significantly different from all others but, in contrast to the MSQ and OSQ depot in growing pigs, IL-6 is not different from most other IL percentages [Hausman et al., unpubl. observation]. Furthermore, IL-5 and IL-12A percentages are significantly different from other IL percentages in neonatal adipose tissue but not in OSQ and MSQ depots [9, 11]. Patterns of TGF gene percentages are clearly age- and depot-dependent. For instance, TGF-β_1 and BMP-15 percentages are significantly different from all other TGF percentages from neonatal adipose tissue [11]. In contrast, only a single TGF percentage is distinguished from all other TGF percentages from OSQ, i.e. TGF-β_3 and ISQ, i.e., TGF-β_2 from growing pigs [9].

Linear regression analysis identified significant linear regressions between individual pig gene expression values for several genes, including leptin, adipose tissue fatty acid-binding protein (AFABP) and individual expression values for a number of other genes for 90-, 150- and 210-day data combined. Notably, IL-6 and IL-15 are distinguished by linear regression analysis of adipose tissue microarray data from growing pigs. Out of 12 ILs and 4 IFNs, IL-6 is the only IL not significantly associated with leptin and AFABP gene expression. Furthermore, IL-15 is the only IL negatively associated with AFABP expression and positively associated with leptin expression. Genes negatively associated with leptin gene expression and positively associated with AFABP expression included IL-1B, IL-1A and other ILs, IFNs, zinc finger protein ubi-d4 (REQUIEM), ring finger protein 19A (DORFIN), nitric oxide synthase-inducible (NOS2A) and leptin receptor (LEPR). Genes positively associated with leptin gene expression and negatively associated with AFABP expression included adipose tissue genes from growing pigs such as IL-15, growth hormone receptor (GHR), transcription factor 7-like 1 (TCF7L1), general transcription factor 21 (GTF21), Prohibitin 2, tyrosine protein kinase 1 (JAK1), low-density-lipoprotein-related proteins (LRP6), nuclear factor 1 B-type

(NFIB), PPAR (peroxisome proliferator-activated receptor) γ-1, -2, PPARG and 15-oxoprostaglandin reductase.

Biological processes associated with genes associated with either leptin or AFABP expression in adipose tissue were studied by analyzing for overrepresentation in GO Biological Process categories followed by clustering of the resulting overrepresented terms. None of the clusters exceeded a cluster score of 3 indicating a weak association between regulatory, non-secreted genes associated with leptin. Significant but negative linear regressions were detected between expression values for leptin and values for IL-1B, IL-1A, IL-4, IL-10, IFN-γ and IFN-B1. Cluster and pathway analysis revealed two significant clusters (4.2, 3.9) indicating an association between these cytokines and leptin expression. In particular, one cluster involved genes involved in chemotaxis, cell death, apoptosis and death response among others (table 2).

Adipose Depot Innervation

Morphological studies revealed that adipose tissue is innervated by adrenergic nerve fibers [12]. Immunocytochemical data revealed that most of the subpopulations of the adrenergic immunoreactive LEPR (LEPR-IR) neurons supplying fat tissue in the pig were positive for NPY and tyrosine hydroxylase immunoactivity [13]. Moreover, immunopositive neurons for LEPR were located in the paraventricular nucleus, ventromedial nucleus, anterior hypothalamic area, preoptic area, arcuate nucleus and supraoptic nucleus [14]. These studies provide the first morphological data demonstrating that hypothalamic LEPR-containing neurons are transsynaptically connected to the perirenal fat depot. Neurons which express LEPR RNA are also located in hypothalamic areas involved in regulating LH [15] and GH secretion [16]. Therefore, the above evidence supports a direct link between hypothalamic neurons in the regulation of fat metabolism and reproduction.

Adipokines and Reproductive Function

Leptin
It is well established that reproductive function is metabolically gated. However, the mechanisms whereby energy stores and metabolic cues influence fertility are yet to be completely understood. The effects of leptin appear to be mediated through modulation of hypothalamic NPY expression [17]. In the pig, presence of biologically-active LEPR in the hypothalamus and pituitary [6] and the fact that leptin increased LH secretion from pig pituitary cells and gonadotropin-releasing hormone (GnRH) release from hypothalamic tissue in vitro [2] suggests that leptin

Table 2. Clusters calculated by DAVID's functional annotation tool with the corresponding GO Biological Process terms. Genes used are the human homolog of porcine genes. Simple regression has shown a negative correlation to leptin in all genes used to create clusters. Genes used IL-1B, IL-1A, IL-4, IL-10, IFNG, and IFNB1

Cluster Themes [ES]	GO terms	GO IDs	p value
Chemotaxis [4.16]	chemotaxis	0006935	4.20E-08
	taxis	0042330	4.20E-08
	defense response	0006952	7.40E-08
	regulation of translation	0006417	7.90E-08
	regulation of cellular biosynthetic process	0031326	1.10E-07
	locomotory behavior	0007626	1.30E-07
	regulation of biosynthetic process	0009889	1.60E-07
	apoptosis	0006915	3.40E-07
	programmed cell death	0012501	3.60E-07
	positive regulation of translation	0045727	4.30E-07
	cell death	0008219	4.70E-07
	death	0016265	4.70E-07
	positive regulation of cellular biosynthetic process	0031328	6.20E-07
	regulation of protein metabolic process	0051246	7.70E-07
	immune response	0006955	8.40E-07
	positive regulation of biosynthetic process	0009891	9.80E-07
	behavior	0007610	1.00E-06
	cytokine biosynthetic process	0042089	1.10E-06
	cytokine metabolic process	0042107	1.20E-06
	positive regulation of protein metabolic process	0051247	2.10E-06
	immune system process	0002376	2.50E-06
	cell development	0048468	3.50E-06
	cytokine production	0001816	4.70E-06
	regulation of cell proliferation	0042127	5.40E-06

acts through the hypothalamus. There is strong evidence from co-localization of LEPR mRNA with NPY gene expression that hypothalamic NPY is the primary potential target for leptin in the pig [13]. Moreover, central administration of NPY suppressed LH secretion and stimulated feed intake and reversed the inhibitory action of leptin on feed intake [18]. However, NPY alone may not mediate the action of leptin, since leptin failed to effect NPY release from pig hypothalamic-preoptic area tissue fragments [2]. Furthermore, metabolic signals may in part be communicated to GnRH neurons via other neuropeptides such as galanin-like peptide [19], α-melanocyte-stimulating hormone (α-MSH) [20], β-endorphin [21], or kisspeptin [22, 23].

The kisspeptins (KiSS) are potent stimulators of the GnRH/LH axis [22–24]. The kisspeptins are a group of structurally related peptides that are products of the KiSS-1 gene [25, 26]. Synthesized as a pre-prohormone, it is cleaved to liberate a 54 amino acid peptide which can be proteolytically processed [27] to shorter variants; all of which share the same amidated C-terminus and retain full biological activity. Kisspeptin acting through its cognate receptor, GPR54, is thought to be an important determinate in the onset of puberty [28]. Thus, hypothalamic KiSS and its receptor GPR54 system may serve as an essential gatekeeper of GnRH neurons and, hence, of reproductive function. Recently, Lents et al. [29] reported that ICV administration of KiSS stimulated LH and follicle-stimulating hormone (FSH) secretion in the prepubertal gilt. Furthermore, the porcine KiSS-1 gene has been cloned and in situ hybridization demonstrated KiSS-1 gene expression in the porcine hypothalamus [Lents et al., unpubl. data]. These data illustrate that KiSS can activate the brain-pituitary axis and may be part of an important mechanism regulating activation of the GnRH/LH axis. In addition, the hypothalamic KiSS-1 gene may convey the modulatory action of metabolic signals to the GnRH neurons [23, 30]. In the mouse, short-term fasting reduced hypothalamic KiSS-1 and GPR54 mRNA levels at 12 and 24 h which preceded the reduction in GnRH gene expression at 48 h [23]. Moreover, leptin stimulated KiSS-1 expression in mouse hypothalamic cell line N6 [23]. Furthermore, hypothalamic KiSS-1 expression was decreased in NPY null mice and this was reversed by NPY administration [23]. These reports support the idea that leptin and NPY are key mediators of metabolic regulation of the hypothalamic KiSS-1 system and subsequent GnRH release. Thus, the evidence supports the notion that the KiSS neuronal pathway may play a role in interfacing metabolic status with the gonadotropic axis in the pig. However, further work is needed to support this hypothesis.

Leptin receptors have been observed in both the granulosa and theca cells of the human [31, 32], bovine [33, 34] and porcine [35]. Circulating concentrations of leptin are generally <10 ng/ml [36–38] and both physiological and supraphysiological leptin concentrations have been reported to effect steroid synthesis in vitro and in vivo. In vitro studies demonstrated that treatment with supraphysiological concentrations of leptin inhibited steroidogenesis in bovine granulosa [33, 39] and theca cells [34]. Similar results have been reported for sheep [40] and pigs [41]. In contrast, physiological doses of leptin have been reported to stimulate steroidogenesis in pig granulosa cells in culture in the presence or absence of IGF-I [41]. Furthermore, leptins' stimulatory influence on steroidogenesis in whole porcine follicle culture by modulating the action of GH and IGF-I or gondotrophins were dependent on follicular maturation [42, 43]. Taken together, the above reports support the idea that leptin has a direct role in modulating follicular development. A critical blood level of leptin may be necessary to initiate follicular development, a stimulatory threshold, and elevated circulating leptin concentrations may reach a putative inhibitory threshold.

The fact that exposure of the somatic cells of the preovulatory follicle to leptin enhanced the steroidogenic capacity of the luteal cells suggests a role for leptin in corpus luteum development. In support of this idea, leptin treatment increased progesterone production from porcine granulosa cells was associated with enhanced expression of steroidogenic acute regulatory protein (StAR), which may be a key regulatory event in the action of leptin on steroidogensis [41]. Moreover, LPR mRNA abundance increased during in vitro luteinization and was greatest in luteal tissue collected during the mid-luteal phase in the pig [35]. Thus, evidence has been reviewed demonstrating that leptin plays role in both follicular development and subsequent luteal function.

Adiponectin
Adiponectin, the most abundantly synthesized protein in adipose tissue, has direct plieotropic effects on ovulation, steroidogenesis and placenta function. Synthesis and expression decrease with increasing adiposity [44], acute food deprivation and fasting. In the pig, adiponectin receptor 1 (ADIPOR1) is primarily expressed in skeletal muscle and adipose tissue, whereas ADIPOR2 was expressed predominantly in liver, heart, skeletal muscle adipose tissue and uterus [45]. Furthermore, recent studies have demonstrated that ADIPOR-1–2 mRNA is localized in the pig, rat and chicken granulosa cells [46–48]. In chickens, ADIPOR-1–2 mRNA abundance was greater in granulosa cells than theca cells [48] but greater in theca cells than granulosa cells of rats [47]. In the bovine, ADIPOR2 mRNA abundance was greater in large follicles than small follicle theca cells [49]. Moreover, adiponectin inhibited progesterone and androstenedione production from theca cells from large follicles induced by LH. In contrast, in the rat granulosa cells, adiponectin had no effect on basal or FSH-induced progesterone or estradiol production and adiponectin inhibited LH-receptor expression in granulosa cells [49]. Ledoux et al. [46] reported that adiponectin acted directly on porcine granulosa cells in vitro inducing expression of genes and proteins similar to the preovulatory process in an ovarian follicle such as upregulation of the cyclooxgenase (COX)-2 and prostaglandin E (PGE) synthase gene. Moreover, adiponectin treatment upregulated StAR and vascular endothelial growth factor (VEGF) and downregulation of cytochrome P450 aromatase (CYP19) expression; these changes are characteristic of those that occur between initiation of the ovulatory process by gonadotropins and subsequent ovulation [46]. These studies suggest species difference in the potential action of adiponectin on follicular function but more importantly provide a potential link with adipose tissue/metabolic state and ovarian function.

Nesfatin

Oh et al. [50] identified nesfatin as a satiety factor, corresponding to NEFA/nucle-obindin-2 gene (NUCB2) as a 396 amino acid product highly conserved between humans and rodents and his expressed in neural and adipose tissue. NUCB2 is post-translationally processed to yield at least three protein fragments termed nes-fatin-1, -2, and -3. Nesfatin-2 and -3 have structures similar to that of DNA or calcium-binding proteins [51]. Nesfatin-1 is comprised of the first 82 amino acids of the proprotein and has a signal sequence, which indicates it is a secreted factor. Nesfatin-1 is present in the peripheral circulation and in the cerebrospinal fluid [50]. A recent report demonstrated that nesfatin-1 was secreted into culture media by human subcutaneous adipose tissue explants and secretion was regulated by the pro-inflammatory cytokines TNFα and IL-6 [52]. Moreover, both NUCB2 mRNA and nafatin-1 protein are expressed in an adipose tissue in a depot-dependent manner with expression being greater in subcutaneous adipose tissue as compared to omental adipose tissue in mice and humans [52]. Immunoreactive and NUCB2 mRNA cell bodies were located in several hypothalamic nuclei, but only in the paraventricular nucleus was expression reduced due to fasting [50]. Intracerebroventricular administration of nesfatin-1 decreased feed intake in a dose-dependent manner whereas ICV administration of nesfatin-1 antibody stimulated appetite in the rat. Furthermore, chronic infusion of anti-nasfatin-1 antibody decreased body weight and fat pad weights in rats [50]. The suppressive effects of nesfatin-1 are independent of leptin, since anti-nesfatin-1 antibody does not block the leptin-induced anorexia [50]. Moreover, central administration of α-MSH enhanced NUCB2 gene expression in the PVN and the satiety effect of nesfatin-1 is blocked by an antagonist to melanocortin receptor (MCR)-3/4, SHU9119 [50]. Thus, it appears that nesfatin-1 interacts with the melanocortin system, a key neural pathway in the regulation of appetite. This is of particular interest, since Rothschild and co-workers identified a missense mutation (D298N) in the porcine MC4R (pMC4R) gene that changes a highly conserved Asp in TM7 in MCR to Asn [53]. Moreover, this missense mutation is associated with increased feed intake, growth, and fatness [53]. A structure and function relationship has also been observed in natural and experimentally induced MC4R mutations in humans [54, 55] and mice [56]. A recent report by Li et al. [57] demonstrated that SNPs of the bovine NUCB2 gene were associated with body weight, body length, heart girth and average daily gain in three Chinese cattle breeds. Furthermore, work in our laboratory demonstrated that ICV injection of [Nle^4, D-Phe^7]-α-MSH, a MC3/4R agonist, decreased NUCB2 expression by 1.5-fold (p = 0.05) in subcutaneous fat in the pig [Barb et al., unpubl. data]. Moreover, ICV administration of α-MSH decreased feed intake and resulted in differential expression of 5,066 genes in adipose and 249 genes in liver tissue that were associated with energy, lipid and carbohydrate metabolism [58]. These changes are similar to expression patterns

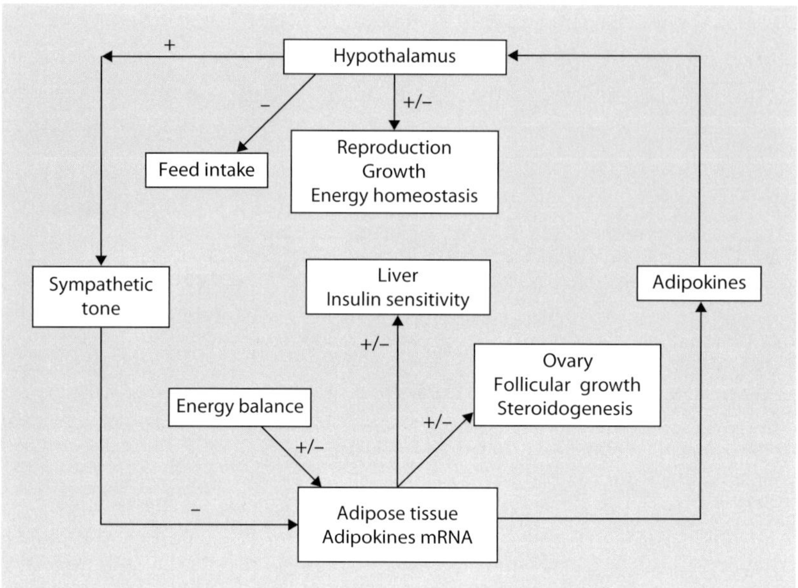

Fig. 1. Schematic illustrating the proposed role of adipokines, e.g. leptin, adiponectin and nesfatin, as important regulators of reproductive function in response to changes in energy balance. Leptin, CNTF and nesfatin have been shown to enhance LH secretion and inhibit appetite. In addition, leptin acts to increase sympathetic tone and thereby regulate adipocye function. A number of these proteins, leptin, adiponectin, and TNFα directly effect ovarian function whereas angiopoietin-like protein-4, resistin and visfatin regulate insulin sensitivity, glucose homeostasis and lipid metabolism (see table 1).

observed in the fasted pig [59]. It should be noted that a recent report demonstrated that nesfatin-1 treatment stimulated LH release in rats [60]. Collectively, these reports suggest that nesfatin-1 may play a role integrating metabolic status with the neuroendocrine axis.

In conclusion, evidence presented supports the concept of the endocrine role of adipose tissue via expression of proteins 'adipokines'. These proteins are regulated by autocrine, endocrine, metabolic and (or) neural signals, which reflect changes in energy balance. A number of these proteins, leptin, CNTF and nesfatin, have been shown to enhance LH secretion and inhibit appetite. A preponderance of evidence cited above suggests that leptin may serve as the primary metabolic signal interacting with neuropeptides such as kisspeptin and NPY that link energy status with the neuroendocrine axis and subsequent reproduction. However, a number of these adipokines have exhibited measureable effects on gonadal function. The best studied of these, leptin and adiponectin, have a detectable effect on steroidogenesis, follicular development and ovulation. Changes in fat metabolism in response to changes in feed intake and energy balance alter adipocyte function and circulating concentrations of adipokines. Thus, a case can be made that adipokines are important regulators of reproductive function during changes in energy balance (fig. 1).

References

1 Barb CR, Hausman GJ, Houseknecht KL: Biology of leptin in the pig. Domest Anim Endocrinol 2001;21:297–317.
2 Barb CR, Barrett JB, Kraeling RR: Role of leptin in modulating the hypothalamic-pituitary axis in the pig. Domest Anim Endocrinol 2004;26:201–214.
3 Campfield LA, Smith FJ, Guisez Y, Devos R, Burn P: Recombinant mouse ob protein: evidence for a peripheral signal linking adiposity and the central neural networks. Science 1995;269:546–549.
4 Zhang Y, Proenca R, Maffel M, Barone M, Leopold L, Friedman JM: Positional cloning of the mouse obese gene and its human homologue. Nature 1994;372:425–432.
5 Dyer CJ, Simmons JM, Matteri RL, Keisler DH: Leptin receptor mRNA is expressed in ewe anterior pituitary and adipose tissue and is differentially expressed in hypothalamic regions of well-fed and feed-restricted ewes. Domest Anim Endocrinol 1997;14:119–128.
6 Lin J, Barb CR, Matteri RL, Kraeling RR, Chen X, Meinersmann RJ, Rampacek GB: Long form leptin receptor mRNA expression in the brain, pituitary, and other tissues in the pig. Domest Anim Endocrinol 2000;19:53–61.
7 Houseknecht KL, Portocarrero CP: Leptin and its receptors: regulators of whole-body energy homeostasis. Domest Anim Endocrinol 1998;15:457–475.
8 Kennedy GC: The role of depot fat in the hypothalamic control of food intake in the rat. Proc Roy Soc Lond B Biol Sci 1953;140:578–592.
9 Hausman GJ, Barb CR, Dean RG: Patterns of gene expression in pig adipose tissue: transforming growth factors, interferons, interleukins and other cytokines. J Anim Sci 2007;85:2445–2456.
10 Hausman GJ, Barb CR, Dean RG: Patterns of gene expression in pig adipose tissue: insulin-like growth factor system proteins, neuropeptide Y (NPY), NPY receptors, neurotrophic factors and other secreted factors. Domest Anim Endocrinol 2008;35:24–34.
11 Hausman GJ, Poulos SP, Richardson RL, Barb CR, Andacht T, Mynatt R, Kirk H: Secreted proteins and genes in fetal and neonatal pig adipose tissue and stromal-vascular cells. J Anim Sci 2006;84:1666–1681.
12 Hausman GJ, Richardson RL: Adrenergic innervation of fetal pig adipose tissue. Histochemical and ultrastructural studies. Acta Anat (Basel) 1987;130:291–297.
13 Czaja K, Lakomy M, Sienkiewicz W, Kaleczyc J, Pidsudko Z, Barb CR, Rampacek GB, Kraeling RR: Distribution of neurons containing leptin receptors in the hypothalamus of the pig. Biochem Biophys Res Commun 2002;298:333–337.
14 Czaja K, Kraeling RR, Barb CR: Are hypothalamic neurons transsynaptically connected to porcine adipose tissue? Biochem Biophys Res Commun 2003;311:482–485.
15 Kineman RD, Leshin LS, Crim JW, Rampacek GB, Kraeling RR: Localization of luteinizing hormone-releasing hormone in the forebrain of the pig. Biol Reprod 1988;39:665–672.
16 Leshin LS, Barb CR, Kiser TE, Rampacek GB, Kraeling RR: Growth hormone-releasing hormone and somatostatin neurons within the porcine and bovine hypothalamus. Neuroendocrinology 1994;59:251–264.
17 Campfield LA, Smith FJ, Burn P: The OB protein (leptin) pathway – a link between adipose tissue mass and central neural networks. Horm Metab Res 1996;28:619–632.
18 Barb CR, Kraeling RR, Rampacek GB, Hausman GJ: The role of neuropeptide Y and interaction with leptin in regulating feed intake and luteinizing hormone and growth hormone secretion in the pig. Reproduction 2006;131:1127–1135.
19 Rich N, Reyes P, Reap L, Goswami R, Fraley GS: Sex differences in the effect of prepubertal GALP infusion on growth, metabolism and LH secretion. Physiol Behav 2007;92:814–823.
20 Crown A, Clifton DK, Steiner RA: Neuropeptide signaling in the integration of metabolism and reproduction. Neuroendocrinology 2007;86:175–182.

21 Barb CR, Chang W-C, Leshin LS, Rampacek GB, Kraeling RR: Opioid modulation of gonadotropin-releasing hormone release from the hypothalamic preoptic area in the pig. Domest Anim Endocrinol 1994;11:375–382.

22 Arreguin-Arevalo JA, Lents CA, Farmerie TA, Nett TM, Clay CM: KiSS-1 peptide induces release of LH by a direct effect on the hypothalamus of ovariectomized ewes. Anim Reprod Sci 2007;101:265–275.

23 Luque RM, Kinemam RD, Tena-Sempere M: Regulation of hypothalamic expression of KiSS-1 and GPR54 genes by metabolic factors: analyses using mouse models and a cell line. Endocrinology 2007;148:4601–4611.

24 Castellano JM, Navarro VM, Fernandez-Fernandez R, Castano JP, Malagon MM, Aguilar E, Dieguez C, Magni P, Pinilla L, Tena-Sempere M: Ontogeny and mechanisms of action for the stimulatory effect of kisspeptin on gonadotropin-releasing hormone system of the rat. Mol Cell Endocrinol 2006;257–258:75–83.

25 Ohtaki T, Shintani Y, Honda S, Matsumoto H, Hori A, Kanehashi K, Terao Y, Kumano S, Takatsu Y, Masuda Y, Ishibashi Y, Watanabe T, Asada M, Yamada T, Suenaga M, Kitada C, Usuki S, Kurokawa T, Onda H, Nishimura O, Fujino M: Metastasis suppressor gene KiSS-1 encodes peptide ligand of a G-protein-coupled receptor. Nature 2001;411:613–617.

26 Kotani M, Detheux M, Vandenbogaerde A, Communi D, Vanderwinden JM, Le Poul E, Brezillon S, Tyldesley R, Suarez-Huerta N, Vandeput F, Blanpain C, Schiffmann SN, Vassart G, Parmentier M: The metastasis suppressor gene KiSS-1 encodes kisspeptins, the natural ligands of the orphan G protein-coupled receptor GPR54. J Biol Chem 2001;276:34631–34636.

27 Takino T, Koshikawa N, Miyamori H, Tanaka M, Sasaki T, Okada Y, Seiki M, Sato H: Cleavage of metastasis suppressor gene product KiSS-1 protein/metastin by matrix metalloproteinases. Oncogene 2003;22:4617–4626.

28 Smith JT, Clarke IJ: Kisspeptin expression in the brain: catalyst for the initiation of puberty. Rev Endocr Metab Disord 2007;8:1–9.

29 Lents CA, Heidorn NL, Barb CR, Ford JJ: Central and peripheral administration of kisspeptin activates gonadotropin but not somatotropin secretion in prepubertal gilts. Reproduction 2008;135:879–887.

30 Tena-Sempere M: KiSS-1 and reproduction: focus on its role in the metabolic regulation of fertility. Neuroendocrinology 2006;83:275–281.

31 Cioffi JA, Van Blerkom J, Antczak M, Shafer A, Wittmer S, Snodgrass HR: The expression of leptin and its receptors in pre-ovulatory human follicles. Mol Hum Reprod 1997;3:467–472.

32 Karlsson C, Lindell K, Svensson E, Bergh C, Lind P, Billig H, Carlsson LMS, Carlsson B: Expression of functional leptin receptors in the human ovary. J Clin Endocrinol Metab 1997;82:4144–4148.

33 Spicer LJ, Francisco CC: The adipose obese gene product, leptin: evidence of a direct inhibitory role in ovarian function. Endocrinology 1997;138:3374–3379.

34 Spicer LJ, Francisco CC: Adipose obese gene product, leptin, inhibits bovine ovarian thecal cell steroidogenesis. Biol Reprod 1998;58:207–212.

35 Ruiz-Cortes ZT, Men T, Palin MF, Downey BR, Lacroix DA, Murphy BD: Porcine leptin receptor: molecular structure and expression in the ovary. Mol Reprod Dev 2000;56:465–474.

36 Delavaud C, Bocquier F, Chilliard Y, Keisler DH, Gertler A, Kann G: Plasma leptin determination in ruminants: effect of nutritional status and body fatness on plasma leptin concentration assessed by a specific RIA in sheep. J Endocrinol 2000;165:519–526.

37 Barb CR, Barrett JB, Kraeling RR, Rampacek GB: Serum leptin concentrations, luteinizing hormone and growth hormone secretion during feed and metabolic fuel restriction in the prepuberal gilt. Domest Anim Endocrinol 2001;20:47–63.

38 Leon HV, Hernandez-Ceron J, Keislert DH, Gutierrez CG: Plasma concentrations of leptin, insulin-like growth factor-I, and insulin in relation to changes in body condition score in heifers. J Anim Sci 2004;82:445–451.

39 Spicer LJ, Chamberlain CS, Francisco CC: Ovarian action of leptin: effects on insulin-like growth factor-I-stimulated function of granulosa and thecal cells. Endocrine 2000;12:53–59.

40 Campbell BK, Guitierrez CG, Armstrong DG, Webb R, Baird DT: Leptin: in vitro and in vivo evidence for direct effects on the ovary in monovulatory ruminants. Hum Reprod 2000;14:15–16.

41 Ruiz-Cortes ZT, Martel-Kennes Y, Gevry NY, Downey BR, Palin MF, Murphy BD: Biphasic effects of leptin in porcine granulosa cells. Biol Reprod 2003;68:789–796.

42 Gregoraszczuk EL, Wojtowicz AK, Ptak A, Nowak K: In vitro effect of leptin on steroids' secretion by FSH- and LH-treated porcine small, medium and large preovulatory follicles. Reprod Biol 2003;3:227–239.

43 Gregoraszczuk EL, Ptak A, Wojtowicz AK, Gorska T, Nowak KW: Estrus cycle-dependent action of leptin on basal and GH or IGF-I stimulated steroid secretion by whole porcine follicles. Endocr Regul 2004;38:15–21.

44 Kadowaki T, Yamauchi T: Adiponectin and adiponectin receptors. Endocr Rev 2005;26:439–451.

45 Lord E, Ledoux S, Murphy BD, Beaudry D, Palin MF: Expression of adiponectin and its receptors in swine. J Anim Sci 2005;83:565–578.

46 Ledoux S, Campos DB, Lopes FL, Dobias-Goff M, Palin MF, Murphy BD: Adiponectin induces periovulatory changes in ovarian follicular cells. Endocrinology 2006;147:5178–5186.

47 Chabrolle C, Tosca L, Dupont J: Regulation of adiponectin and its receptors in rat ovary by human chorionic gonadotrophin treatment and potential involvement of adiponectin in granulosa cell steroidogenesis. Reproduction 2007;133:719–731.

48 Chabrolle C, Tosca L, Crochet S, Tesseraud S, Dupont J: Expression of adiponectin and its receptors (AdipoR1 and AdipoR2) in chicken ovary: potential role in ovarian steroidogenesis. Domest Anim Endocrinol 2007;33:480–487.

49 Lagaly DV, Aad PY, Grado-Ahuir JA, Hulsey LB, Spicer LJ: Role of adiponectin in regulating ovarian theca and granulosa cell function. Mol Cell Endocrinol 2008;284:38–45.

50 Oh I, Shimizu H, Satoh T, Okada S, Adachi S, Inoue K, Eguchi H, Yamamoto M, Imaki T, Hashimoto K, Tsuchiya T, Monden T, Horiguchi K, Yamada M, Mori M: Identification of nesfatin-1 as a satiety molecule in the hypothalamus. Nature 2006;443:709–712.

51 Gonzalez R, Tiwari A, Unniappan S: Pancreatic β cells colocalize insulin and pronesfatin immunoreactivity in rodents. Biochem Biophys Res Commun 2009;381:643–648.

52 Ramanjaneya M, Chen J, Brown JEP, Tan BK, Patel S, Lehenert H, Randrva HS: Identification of a novel adipokine nesfatin-1/NUCB2 in human and murine adipose tissue: altered levels in obesity and food deprivation. Proc Endocrine Society Annual Meeting, OR31-1, 2009.

53 Kim KS, Larsen N, Short T, Plastow G, Rothschild MF: A missense variant of the porcine melanocortin-4 receptor (MC4R) gene is associated with fatness, growth, and feed intake traits. Mamm Genome 2000;11:131–135.

54 Vaisse C, Clement K, Durand E, Hercberg S, Guy-Grand B, Froguel P: Melanocortin-4 receptor mutations are a frequent and heterogeneous cause of morbid obesity. J Clin Invest 2000;106:253–262.

55 Lubrano-Berthelier C, Cavazos M, Le Stunff C, Haas K, Shapiro A, Zhang S, Bougneres P, Vaisse C: The human MC4R promoter: characterization and role in obesity. Diabetes 2003;52:2996–3000.

56 Meehan TP, Tabeta K, Du X, Woodward LS, Firozi K, Beutler B, Justice MJ: Point mutations in the melanocortin-4 receptor cause variable obesity in mice. Mamm Genome 2006;17:1162–1171.

57 Li F, Chen H, Lei CZ, Ren G, Wang J, Li ZJ, Wang JQ: Novel SNPs of the bovine NUCB2 gene and their association with growth traits in three native Chinese cattle breeds. Mol Biol Rep 2010;37:501–505.

58 Barb CR, Hausman GJ, Rekaya R, Lents CA, Lkhagvadorj S, Qu L, Cai W, Couture OP, Anderson LL, Dekkers JCM, Tuggle CK: Microarray gene expression profiles in hypothalamus, liver and adipose tissues and feed intake response to melanocortin-4 receptor agonist in pigs expressing melanocortin-4 receptor mutations. Physiol Genomics 2009.

59 Lkhagvadorj S, Qu L, Cai W, Couture OP, Barb CR, Hausman GJ, Nettleton D, Anderson LL, Dekkers JCM, Tuggle CK: Microarray gene expression profiles of fasting induced changes in liver and adipose tissues of pigs expressing the melanocortin-4 receptor D298N variant. Physiol Genomics 2009;38:98–111.

60 Garcia-Galiano D, Navarro VM, Sanchez-Garrido MA, Pineda R, Castellano JM, Romero M, Aguilar E, Dieguez C, Pinilla L, Tena-Sempere M: The anorexigenic neuropeptide, nesfatin-1, as putative regulator of the gonadotropic axis and puberty onset in the female rat. Proc Endocrine Society Annual Meeting, OR31-1, Washington 2009.

61 Fukuhara A, Matsuda M, Nishizawa M, Segawa K, Tanaka M, Kishimoto K, Matsuki Y, Murakami M, Ichisaka T, Murakami H, Watanabe E, Takagi T, Akiyoshi M, Ohtsubo T, Kihara S, Yamashita S, Makishima M, Funahashi T, Yamanaka S, Hiramatsu R, Matsuzawa Y, Shimomura I: Visfatin: a protein secreted by visceral fat that mimics the effects of insulin. Science 2005;307:426–430.

62 Plati E, Kouskouni E, Malamitsi-Puchner A, Boutsikou M, Kaparos G, Baka S: Visfatin and leptin levels in women with polycystic ovaries undergoing ovarian stimulation. Fertil Steril 2009; doi:10.1016/j.fertnstert.2009.04.055. [Epub ahead of print].

63 Bastard JP, Maachi M, Lagathu C, Kim MJ, Caron M, Vidal H, Capeau J, Feve B: Recent advances in the relationship between obesity, inflammation, and insulin resistance. Eur Cytokine Netw 2006; 17:4–12.

64 Korzekwa A, Murakami S, Woclawek-Potocka I, Bah MM, Okuda K, Skarzynski DJ: The influence of tumor necrosis factor-α (TNF) on the secretory function of bovine corpus luteum: TNF and its receptors expression during the estrous cycle. Reprod Biol 2008;8:245–262.

65 Nagyova E, Nemcova L, Prochazka R: Expression of tumor necrosis factor-α-induced protein-6 messenger RNA in porcine preovulatory ovarian follicles. J Reprod Dev 2009;55:231–235.

66 Gorospe WC, Hughes FM Jr, Spangelo BL: Interleukin-6: effects on and production by rat granulosa cells in vitro. Endocrinology 1992;130: 1750–1752.

67 Gorospe WC, Spangelo BL: Interleukin-6 production by rat granulosa cells in vitro: effects of cytokines, follicle-stimulating hormone, and cyclic 3′,5′-adenosine monophosphate. Biol Reprod 1993;48:538–543.

68 Hughes FM Jr, Lane TA, Chen TT, Gorospe WC: Effects of cytokines on porcine granulosa cell steroidogenesis in vitro. Biol Reprod 1990;43:812–817.

69 Kalra SP, Xu B, Dube MG, Moldwaer LL, Martin D, Kalra PS: Leptin and ciliary neurotropic factor inhibit fasting-induced suppression of luteinizing hormone release in rats: role of neuropeptide Y. Neurosci Lett 1998;240:45–49.

70 Xu B, Dube MG, Kalra PS, Farmerie WG, Kaibara A, Moldawer LL, Martin D, Kalra SP: Anorectic effects of the cytokine, ciliary neurotropic factor, are mediated by hypothalamic neuropeptide Y: comparison with leptin. Endocrinology 1998;139: 466–473.

71 Vacher CM, Crepin D, Aubourg A, Couvreur O, Bailleux V, Nicolas V, Ferezou J, Gripois D, Gertler A, Taouis M: A putative physiological role of hypothalamic CNTF in the control of energy homeostasis. FEBS Lett 2008;582:3832–3838.

72 Sukonina V, Lookene A, Olivecrona T, Olivecrona G: Angiopoietin-like protein-4 converts lipoprotein lipase to inactive monomers and modulates lipase activity in adipose tissue. Proc Natl Acad Sci USA 2006;103:17450–17455.

73 Xu A, Lam MC, Chan KW, Wang Y, Zhang J, Hoo RL, Xu JY, Chen B, Chow WS, Tso AW, Lam KS: Angiopoietin-like protein-4 decreases blood glucose and improves glucose tolerance but induces hyperlipidemia and hepatic steatosis in mice. Proc Natl Acad Sci USA 2005;102:6086–6091.

74 Chen XD, Lei T, Xia T, Gan L, Yang ZQ: Increased expression of resistin and tumour necrosis factor-α in pig adipose tissue as well as effect of feeding treatment on resistin and cAMP pathway. Diabetes Obes Metab 2004;6:271–279.

75 Mitchell M, Armstrong DT, Robker RL, Norman RJ: Adipokines: implications for female fertility and obesity. Reproduction 2005;130:583–597.

76 Otsuka F, Yao Z, Lee T, Yamamoto S, Erickson GF, Shimasaki S: Bone morphogenetic protein-15. Identification of target cells and biological functions. J Biol Chem 2000;275:39523–39528.

77 Otsuka F, Yamamoto S, Erickson GF, Shimasaki S: Bone morphogenetic protein-15 inhibits follicle-stimulating hormone (FSH) action by suppressing FSH receptor expression. J Biol Chem 2001; 276:11387–11392.

78 Otani H, Otsuka F, Takeda M, Mukai T, Terasaka T, Miyoshi T, Inagaki K, Suzuki J, Ogura T, Lawson MA, Makino H: Regulation of GNRH production by estrogen and bone morphogenetic proteins in GT1–7 hypothalamic cells. J Endocrinol 2009;203:87–97.

Gary J. Hausman
USDA, ARS, Richard B. Russell Agriculture Research Center
950 College Station Rd, Athens, GA 30605 (USA)
Tel. +1 706 546 3124, Fax +1 706 546 3633
E-Mail gary.hausman@ars.usda.gov

Levy-Marchal C, Pénicaud L (eds): Adipose Tissue Development: From Animal Models to Clinical Conditions.
Endocr Dev. Basel, Karger, 2010, vol 19, pp 45–52

Unraveling the Obesity and Breast Cancer Links: A Role for Cancer-Associated Adipocytes?

Béatrice Dirat[a–c] · Ludivine Bochet[a–c] · Ghislaine Escourrou[d] ·
Philippe Valet[a,c] · Catherine Muller[a,b]

[a]University of Toulouse, [b]Institute of Pharmacology and Structural Biology CNRS UMR 5089, [c]Institut
National de la Santé et de la Recherche médicale, INSERM U858 Equipe 3, IFR31, and [d]Department of
Anatompathology and Cytology, CHU Rangueil, Toulouse, France

Abstract

In addition to diabetes and cardiovascular diseases, epidemiological evidence demonstrates
that people who are obese or overweight are at increased risk of developing cancer – colon,
breast (in postmenopausal women), endometrial or kidney cancer being among the most frequent. In addition to the increase in tumor occurrence, obesity also affects tumor prognosis,
especially in breast and prostate cancers. In breast cancer, obesity is associated with reduced
survival and increased recurrence independent of menopausal status. Host factors seem to contribute to the occurrence of tumors exhibiting an aggressive biology defined by advanced stages
and high grade. Mature adipocytes are part of the breast cancer tissue and as highly endocrine
cells susceptible to profoundly modify breast cancer cell behavior. Tumor progression has
recently been recognized as the product of an evolving crosstalk between tumor cells and the
surrounding 'normal' cells. We propose that such a bidirectional crosstalk exists between breast
cancer cells and tumor-surrounding adipocytes, and that the tumor-modified adipocytes (or
cancer-associated adipocytes) are key actors in tumor progression. The positive contribution of
cancer-associated adipocytes into tumor progression might be amplified in obese women and
explains at least in part the poor prognosis observed in this subset of patients.

The prevalence of overweight and obesity has been increasing worldwide over the
past decades and reached alarming dimensions [1]. According to 2005 projections
by the World Health Organization (WHO), there are 1.6 billion overweight and
400 million obese adults worldwide. WHO predicts that by 2015 these numbers
will increase to 2.3 billion overweight and 700 million obese adults Studies from

different countries with different lifestyles have shown consistently that obesity, as measured using body mass index (BMI), is associated with an increased risk of all-cause mortality [1, 2]. In studies able to stratify the cause of death, the increased all-cause mortality has been related to cardiovascular and cancer mortality [see e.g. 2]. Although obesity has long been recognized as an important cause of diabetes and cardiovascular disease, the relationship between obesity and cancer has retained less attention than its cardiovascular effects. In a recent interview, Prof. D. Hill, President of the UICC, said 'Lack of public understanding of the link between body weight and cancer probably parallels our attitudes to smoking and cancer in the late 1950s'. However, the link between obesity and cancer is complex since cancer is a heterogeneous group of diseases. The International Agency on Cancer has determined that people who are overweight or obese are at increased risk of developing adenocarcinoma of the esophagus, colon cancer, breast cancer (in postmenopausal women), endometrial cancer and kidney cancer. Some epidemiological evidence has also indicated that cancers of the liver, gallbladder and pancreas are also obesity-related whereas no associations are seen for example for lung cancer [reviewed in 2]. In addition, obesity might influence different steps of the disease and a careful examination should be given when considering the diagnosis (increase in cancer occurrence) or prognosis and mortality. For example, in breast cancer, early studies have established that obese women have an increased risk of developing postmenopausal, but not premenopausal, breast cancer. It has been even demonstrated that a modest reduction in risk is present in women with high BMI [reviewed in 2]. By contrast, studies of mortality and survival demonstrated that obesity is associated with reduced survival and an increased likelihood of recurrence. This poor prognosis seems to be independent of menopausal status, smoking habits, tumor stage, and tumor hormone-binding characteristics [for reviews, see 3, 4]. Multiple factors might contribute to the survival rates in obese patients with cancer including a higher likelihood of comorbid conditions and other 'non-biological' effects (e.g. delayed diagnosis or underdosing of chemotherapy). However, emerging evidence suggests that host factors might contribute to the occurrence in obese women of tumors exhibiting an aggressive biology defined by advanced stage and high grade [5–8]. The molecular mechanisms underlying these effects remain poorly understood but increasing interest has emerged towards adipokines, a term that defined the multiple secretory products of adipocytes. In fact, adipocytes, initially considered as energy storage cells, have clearly emerged as endocrine cells in the last 10 years [9]. Through their ability to secrete hormones, growth factors, chemokines or pro-inflammatory molecules, adipocytes are therefore excellent candidates to play a crucial role in tumor behavior through heterotypic signaling processes. The crosstalk between adipose and epithelial tumor components might be positively affected in obesity where the normal balance of these adipose tissue secretory proteins is perturbed [10, 11].

Therefore, the purpose of this review is to emphasize the role that mammary adipose tissue might play in locally affecting breast cancer cell behavior. Adipose tissue constitutes an active endocrine organ that can have far-reaching effects on the physiology of other tissues. To understand how obesity influences breast cancer prognosis by increasing circulating plasma levels of estrogen, insulin, insulin-like growth factors or other hormonal factors, the reader is referred to recent reviews on that topic [see 2, 12].

Adipose Tissue Is an Active Component of the Tumor Stroma

Emerging evidence in the literature recognizes tumor progression as the product of an evolving crosstalk between tumor cells and its surrounding supportive tissue, the tumor stroma [13]. This framework includes a specific type of extracellular matrix – the tumor matrix – as well as cellular components such as fibroblasts, immune and inflammatory cells, and blood vessels. Based on the pioneering work of Jude Folkman [14], it is now widely recognized that tumors must induce angiogenesis to grow as well as to metastasize to distant organ sites and accordingly targeting the tumor vasculature has become an attractive possibility as a form of anticancer therapy. Recent cumulative evidence demonstrates that in addition to endothelial cells, other components of the stroma also act as key players in tumor progression and invasion. Cancer-associated fibroblasts and tumor-associated macrophages promote tumor progression by secreting growth factors, chemokines and pro-migratory extracellular matrix components [13, 15]. Is there a role for adipocytes in this 'tissue-integrated' view of tumor progression? Compared to other components of the stroma, adipocytes received relatively little attention despite the fact that they correspond to one of the most prominent cell types in breast tumors (fig. 1). In addition, an effect of tumor-surrounding adipocytes would not have been entirely surprising when one considers the intimate relationship that exists between adipose and mammary tissues. The mammary gland develops in a bed of white adipose tissue, and there are extensive mesenchymal-epithelial interactions in which the adipose mesenchyme 'instructs' aspects of mammary epithelial development [16]. Finally, it is obvious that in numerous organs including breast, early local tumor invasion results in immediate proximity of cancer cells to adipocytes (fig. 1). There is some evidence in the literature that adipocytes can affect through paracrine mechanisms the behavior of breast tumor cells [for review, see 12]. For example in a three-dimensional collagen gel matrix co-culture technique, it has been shown that mature adipocytes secrete a factor(s) which can promote the growth of human breast cancer cell lines, as reflected in an enhanced uptake of bromodeoxyuridine [17]. In an elegant in vivo study, the group of Scherer [18] has shown that co-injection of the breast cancer cell line SUM59PT with murine

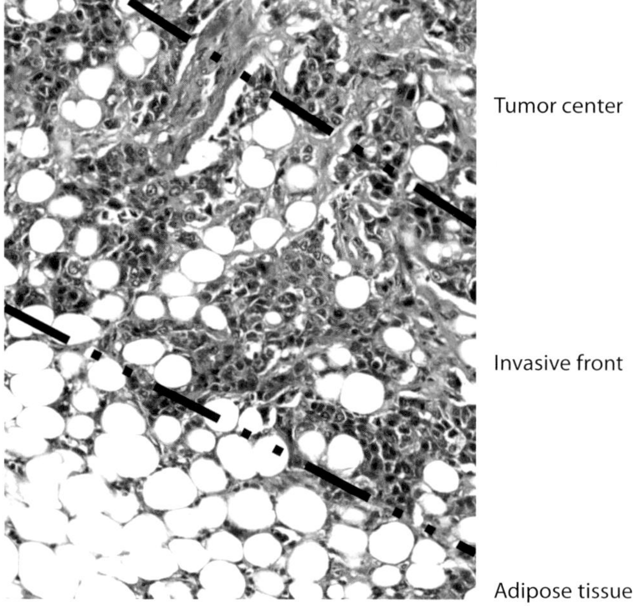

Fig. 1. Mature adipocytes are present in breast tumor stroma. Histologic examination of a breast tumor (HE, orig. magnif. ×200). Note that mature adipocytes are located adjacent to invading cancer cells.

adipocytes (but not pre-adipocytes) in immunodeficient mice not only increase the levels and rate of tumor progression but also promote development of metastases as well. They further demonstrate that this effect is partially controlled by type VI collagen secreted by adipocytes that promotes GSK3β phosphorylation, β-catenin stabilization and increased β-catenin activity in breast cancer cells [19]. Collagen VI is not the only candidate that has been demonstrated to be involved in this process of adipokine's driven tumor progression. Leptin has been shown to promote in some, but not all, cell lines in an autocrine and paracrine manner the proliferation and/or migration of breast cancer cells. Another adipose tissue-derived protein, adiponectin, whose expression is reported to be inversely related to body fat, has been shown to inhibit cell proliferation and enhance breast cancer cell apoptosis, although again these effects appear to be restricted to specific models (for a detailed review of these two adipokines in breast cancer progression, see Vona-Davis and Rose [12]). Elevated serum estrogen levels as well as enhanced local production of estrogen have been considered primary mediators of how increased body weight promotes breast cancer development in postmenopausal women [2]. Related to tumor progression, the local estrogen production (through aromatization of the C19 steroid androstenedione in adipocytes) might also favor tumor growth through the interaction with the estrogen receptor of nearby tumor cells, a point that has been extensively reviewed in the literature [for review, see 20]. In obesity, the normal balance of these adipose tissue secretory proteins is affected both positively (leptin) and negatively (adiponectin). According to the preclinical evidence demonstrating their ability to modulate breast cancer cell behavior, they represent a good candidate to explain the poorer prognosis of breast cancer in obese patients, although their local secretion in breast tumors remains poorly characterized [21]. However, all these studies through their use of 'naive' adipocyte-conditioned medium or the use of recombinant proteins (e.g. leptin or adiponectin) precludes that the mature adipocytes remain relatively inert in response to the nearby tumor cells. It has been demonstrated that cancer cells usually go about generating a supportive microenvironment by modifying the phenotype of surrounding normal cells. For example, it has been demonstrated that cancer cells modify the phenotype of fibroblasts that acquire traits of wounded fibroblasts (including the expression of a smooth cell actin) [22], that are therefore named carcinoma-associated fibroblasts, supporting the idea that cancer is a 'wound that never heals' [13]. In accordance with this concept of a constant bidirectional crosstalk between cancer and stroma cells, we therefore propose that tumor-surrounding adipocytes also exhibit profound phenotypic changes. A first argument in favor of this hypothesis came from the laboratory of Marie-Christine Rio [23] showing that in vivo invasive cancer cells induce in adjacent adipocytes the expression of stromelysin-3 (also named metalloprotease 11, MMP-11), a potent inhibitor of adipocyte differentiation. We recently set up in the laboratory an original co-culture model between breast tumor cells and adipocytes. Our results show that tumor cells promote extensive phenotypic changes in adipocytes leading to an 'activated' phenotype, that in turns promote the occurrence of metastatic traits in cancer cells [Dirat et al., in preparation]. Therefore, these preliminary results demonstrate that cancer cells might force adipocytes to express a new subset of molecules that are not expressed by adipocytes in normal conditions, and that like cancer-associated fibroblasts, cancer-associated adipocytes (CAA) are present within a tumor with specific features that remain to be extensively characterized. A detailed proteomic analysis of the proteins secreted in breast adipose tissue collected from tumorectomy of estrogen receptor-negative, high-grade tumors present in premenopausal women showed that numerous hormones, cytokines and growth factors are present within this tissue [24]. Comparison with adipose tissue of normal breast will clearly offer the opportunity to characterize the molecular circuitry between adipocytes and epithelial cells. In addition, this crosstalk might also affect less differentiated cells as well as mature adipocytes. For example, it has been demonstrated that conditioned media from breast cancer cells inhibited the differentiation of pre-adipocytes into mature adipocytes in favor of the accumulation of pre-adipocyte cells into the cancer tissue [25]. Characterization of the molecular circuitry between CAA and the epithelial component might help

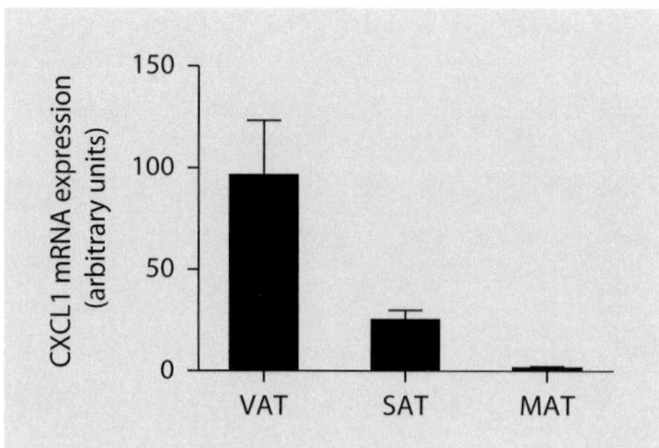

Fig. 2. Chemokine CXCL-1 is not expressed in mammary adipose tissue. Expression of CXCL-1 was measured by qPCR on subcutaneous adipose tissue (SAT). Visceral adipose tissue (VAT) was obtained from healthy donors (5 and 9 respectively) with BMI ≤25. Mammary adipose tissue (MAT) was obtained from reduction mammoplasty performed in 12 healthy women exhibiting normal (≤25) BMI. Results are expressed as mean ± SD (in arbitrary units).

to identify attractive targets in cancer therapy since drugs aimed at stromal adipocyte signals and effector functions may prove complementary to conventional treatments targeting the overt cancer cells in breast tumors.

Consequences for the Poor Prognosis Observed for Breast Cancer in Obese Patients: Some Answers, Many Questions

Although not demonstrated experimentally, one might reasonably speculate that the positive contribution of these CAA into tumor progression might be amplified in obese women, therefore explaining in part the poor prognosis observed in this subset of patients. Accumulating evidence in recent years has demonstrated that obesity is associated with a low-grade inflammation of white adipose tissue which can subsequently lead to insulin resistance, impaired glucose tolerance and even diabetes [9]. The increased production and secretion of a wide range of inflammatory molecules, including TNF-α and interleukin-6, by adipocytes (but also by infiltrating macrophages which may be a major source of locally produced proinflammatory cytokines) seen in white adipose tissue from obese patients might have a profound effect on tumor progression, especially if this inflammatory response is amplified in the cancer tissue [2]. Concerning adipose breast tissue, characterization of its pattern of secretion is clearly needed in both lean and obese patients since it is poorly defined as compared to other fat cell depots. Preliminary experiments performed in our laboratory in fact suggest that striking differences might be observed. For example, as seen in figure 2, whereas the proangiogenic chemokine CXCL-1 is expressed in deep and subcutaneous adipose tissue, its expression is absent from mammary adipose tissue, clearly demonstrating that specific studies should be performed. If, as we believe, the tumor characteristics are modified during the early steps of tumor progression (e.g. early acquisitions of metastatic traits) in a context of obesity, weight loss after breast cancer diagnosis might have a limited impact on prognosis, this latter point remaining largely debated in the literature [26]. Obesity also affects the prognosis of prostate tumors with the occurrence of more aggressive disease [2]. The expected striking increase in the incidence of these aggressive cancers in the near future related to obesity turns the knowledge of this field of great impact as it is needed for the development of strategies to prevent and treat this disease.

Acknowledgements

Studies performed in our laboratories were supported by the French National Cancer Institute (INCA PL 2006–035 to G.E., P.V. and C.M.), the 'Ligue Régionale contre le Cancer' (Comité du Lot et Haute-Garonne to C.M.) and the University of Toulouse (AO CS 2009 to C.M.).

References

1 Yach D, Stuckler D, Brownell KD: Epidemiologic and economic consequences of the global epidemics of obesity and diabetes. Nat Med 2006;12: 62.

2 Calle EE, Kaaks R: Overweight, obesity and cancer: epidemiological evidence and proposed mechanisms. Nat Rev Cancer 2004;4:579.

3 Carmichael AR: Obesity and prognosis of breast cancer. Obes Rev 2006;7:333.

4 Stephenson GD, Rose DP: Breast cancer and obesity: an update. Nutr Cancer 2003;45:1.

5 Feigelson HS, Patel AV, Teras LR, Gansler T, Thun MJ, Calle EE: Adult weight gain and histopathologic characteristics of breast cancer among postmenopausal women. Cancer 2006;107:12.

6 Berclaz G, Li S, Price KN, Coates AS, Castiglione-Gertsch M, Rudenstam, CM, Holmberg SB, Lindtner J, Erien D, Collins J, Snyder R, Thurlimann B, Fey, MF, Mendiola C, Werner ID, Simoncini E, Crivellari D, Gelber D, Goldhirsch A: Body mass index as a prognostic feature in operable breast cancer: the International Breast Cancer Study Group experience. Ann Oncol 2004;15: 875.

7 Daniell HW, Tam E, Filice A: Larger axillary metastases in obese women and smokers with breast cancer – an influence by host factors on early tumor behavior. Breast Cancer Res Treat 1993;25:193.

8 Daling JR, Malone KE, Doody DR, Johnson LG, Gralow JR, Porter PL: Relation of body mass index to tumor markers and survival among young women with invasive ductal breast carcinoma. Cancer 2001;92:720.

9 Rajala MW, Scherer PE: The adipocyte – at the crossroads of energy homeostasis, inflammation, and atherosclerosis. Endocrinology 2003;144: 3765.

10 Chavey C, Boucher J, Monthouel-Kartmann MN, Sage EH, Castan-Laurell I, Valet P, Tartare-Deckert S, Van Obberghen E: Regulation of secreted protein acidic and rich in cysteine during adipose conversion and adipose tissue hyperplasia. Obesity (Silver Spring) 2006;14:1890.

11 Boucher J, Castan-Laurell I, Daviaud D, Guigne C, Buleon M, Carpene C, Saulnier-Blache JS, Valet P: Adipokine expression profile in adipocytes of different mouse models of obesity. Horm Metab Res 2005;37:761.

12 Vona-Davis L, Rose DP: Adipokines as endocrine, paracrine, and autocrine factors in breast cancer risk and progression. Endocr Relat Cancer 2007; 14:189.

13 Mueller MM, Fusenig NE: Friends or foes – bipolar effects of the tumour stroma in cancer. Nat Rev Cancer 2004;4:839.

14 Naumov GN, Akslen LA, Folkman J: Role of angiogenesis in human tumor dormancy: animal models of the angiogenic switch. Cell Cycle 2006; 5:1779.

15 Bissell MJ, Labarge MA: Context, tissue plasticity, and cancer: are tumor stem cells also regulated by the microenvironment? Cancer Cell 2005;7:17.

16 Wiseman BS, Werb Z: Stromal effects on mammary gland development and breast cancer. Science 2002;296:1046.

17 Manabe Y, Toda S, Miyazaki K, Sugihara H: Mature adipocytes, but not preadipocytes, promote the growth of breast carcinoma cells in collagen gel matrix culture through cancer-stromal cell interactions. J Pathol 2003;201:221.

18 Iyengar P, Combs TP, Shah SJ, Gouon-Evans V, Pollard JW, Albanese C, Flanagan L, Tenniswood MP, Guha C, Lisanti MP, Pestell RG, Scherer PE: Adipocyte-secreted factors synergistically promote mammary tumorigenesis through induction of anti-apoptotic transcriptional programs and proto-oncogene stabilization. Oncogene 2003;22: 6408.

19 Iyengar P, Espina V, Williams TW, Lin Y, Berry D, Jelicks LA, Lee H, Temple K, Graves R, Pollard J, Chopra N, Russell RG, Sasisekharan R, Trock BJ, Lippman M, Calvert VS, Petricoin EF 3rd, Liotta L, Dadachova E, Pestell RG, Lisanti, MP, Bonaldo P, Scherer PE: Adipocyte-derived collagen VI affects early mammary tumor progression in vivo, demonstrating a critical interaction in the tumor/ stroma microenvironment. J Clin Invest 2005;115: 1163.

20 Cleary MP, Grossmann ME: Minireview: Obesity and breast cancer: the estrogen connection. Endocrinology 2009;150:2537.

21 Hede K: Fat may fuel breast cancer growth. J Natl Cancer Inst 2008;100:298.

22 Haviv I, Polyak K, Qiu W, Hu M, Campbell I: Origin of carcinoma-associated fibroblasts. Cell Cycle 2008;8:589.

23 Andarawewa KL, Motrescu ER, Chenard MP, Gansmuller A, Stoll I, Tomasetto, C, Rio MC: Stromelysin-3 is a potent negative regulator of adipogenesis participating to cancer cell-adipocyte interaction/crosstalk at the tumor invasive front. Cancer Res 2005;65:10862.

24 Celis JE, Moreira JM, Cabezon T, Gromov P, Friis E, Rank F, Gromova I: Identification of extracellular and intracellular signaling components of the mammary adipose tissue and its interstitial fluid in high-risk breast cancer patients: toward dissecting the molecular circuitry of epithelial-adipocyte stromal cell interactions. Mol Cell Proteomics 2005;4:492.

25 Meng L, Zhou J, Sasano H, Suzuki T, Zeitoun KM, Bulun SE: Tumor necrosis factor-α and interleukin-11 secreted by malignant breast epithelial cells inhibit adipocyte differentiation by selectively down-regulating CCAAT/enhancer binding protein-α and peroxisome proliferator-activated receptor-γ: mechanism of desmoplastic reaction. Cancer Res 2001;61:2250.

26 Griggs JJ, Sabel MS: Obesity and cancer treatment: weighing the evidence. J Clin Oncol 2008; 26, 4060.

Prof. Catherine Muller
IPBS CNRS UMR 5089, Université de Toulouse
205, route de Narbonne, FR–31077 Toulouse Cedex (France)
Tel. +33 5161 17 59 32, Fax +33 561 17 59 33
E-Mail muller@ipbs.fr

 Dirat · Bochet · Escourrou · Valet · Muller

Levy-Marchal C, Pénicaud L (eds): Adipose Tissue Development: From Animal Models to Clinical Conditions.
Endocr Dev. Basel, Karger, 2010, vol 19, pp 53–61

Early Determinants of Obesity

Ken K. Ong

MRC Epidemiology Unit, Cambridge, UK

Abstract

High rates of overweight and obesity even in very young children argue the case for strategies to prevent overweight from very young ages. Historical studies, prospective birth cohorts, and more recently genetic studies all indicate that the rapid weight gain trajectory to later obesity starts in the first months of life, even from birth. Early puberty and age at menarche are consequences of rapid infant weight gain and childhood overweight, and in turn these adolescent traits are predictive for obesity, diabetes, hypertension and cardiovascular disease events in later life. Understanding of the nutritional, parental and wider determinants of rapid infant weight gain are informing the development of obesity prevention strategies starting in early life. Such strategies could be further refined by future studies that address the specific regulation of infant adiposity, and also by studies that explore whether these life-course trajectories are modifiable during adolescence.

Obesity has important consequences for morbidity and mortality during childhood [1] and also during later life. Childhood obesity has been shown to track into adult life [1, 2] and recently Baker et al. [3] reported a linear association between childhood body mass index (BMI) from age 7 to 13 years and the risk of having a coronary heart disease event in adulthood. Currently, there is little evidence for any effective strategy to prevent childhood obesity [4]. The increasing prevalence of overweight and obesity is apparent throughout childhood, and even at very young ages. Routine data in 3-year-olds from the Wirral, England, show that at this age the prevalence of overweight and obesity rose from 12.2 and 2.2% respectively in 1988 to 19.1 and 4.0% in 2003 [5]. Therefore, the prediction and prevention of childhood obesity should include efforts that start very early in life.

Infancy Is a Risk Period for Later Obesity

Studies in the ALSPAC cohort identified a strong positive association between rapidity of infancy weight gain and subsequent obesity risk [6]. Since that original report the association of the rapidity of infant weight gain with obesity risk has been confirmed in many other studies and has been extended into adult life, as summarized in three recent systematic reviews [7–9]. The latest review in 2006 identified 21 studies which had investigated this association, all of which had reported a positive association between infancy weight gain and later BMI or obesity [9].

In addition to outcomes based on BMI, the rapidity of infancy weight gain is positively associated with subsequent adiposity and metabolic risk markers. In the SWEDES study, faster infancy weight between birth to age 6 months was associated with increased percent body fat at the age of 17 years [10], and also with a higher metabolic syndrome risk score [11]. In a case-control study of Spanish children with repeated measures of body composition, low birth weight infants continued to gain excess total body fat and abdominal fat even after completion of rapid catch-up weight gain [12]. Those results were consistent with findings from the German DONALD cohort study which supported the hypothesis that rapid growth during infancy may not only be accompanied by greater concurrent gains in body fat mass, but may also lead to an ongoing propensity to accumulate fat mass and central fat [13]. Studies in the ALSPAC cohort and in low birth weight infants in Chile have reported associations between rapidity of infancy weight gain and other potential metabolic markers of disease risk, including insulin sensitivity and insulin secretion [14], urine steroid profiles [15], IGF-1, GHBP [16] and adiponectin levels [17].

Those observational studies have been supported by other reports of the long-term follow-up of randomized feeding trials in nasogastric tube-fed pre-term infants. In those studies, greater dietary energy content during the first 4 weeks of life had remarkable adverse effects on later body size and metabolic disease risk [18]. Together these studies have supported increasing recognition in government policies of the need to address obesity risks factors from very early on in life, during pregnancy and infancy [19].

Infancy Weight and Subsequent Puberty

In the ALSPAC study, faster infancy weight was associated with earlier age at menarche and also increased adiposity in girls [20]. In mothers from the ALSPAC study, age at menarche was strongly predictive for adult pre-pregnancy obesity risk. There was a fivefold increase in obesity risk across the quintiles of age at menarche [21]. In the EPIC-Norfolk cohort, earlier age at menarche consequently also predicted increased risk for adult-onset diabetes [22]. Recent data from the EPIC-Norfolk cohort identified similar increased risks for hypertension and cardiovascular disease events: even after adjustment for BMI and waist circumference, women who had early menarche (<12 years) had higher risks of hypertension (1.13 [1.02–1.24]), incident CVD (1.17 [1.07–1.27]), incident coronary heart disease (1.23 [1.06–1.43]), all-cause mortality (1.22 [1.07–1.39]), CVD mortality (1.28 [1.02–1.62]) and cancer mortality (1.25 [1.03–1.51]) compared to women with later menarche [23].

As well as predicting later disease in adults, rapid infancy weight gain particularly following low birth weight may have adverse consequences for adolescent health outcomes, including precocious pubarche, polycystic ovary syndrome, and shorter final height [24–26]. Those studies have also identified that the timing of puberty may indeed be mutable with simple interventions; early completion of puberty was prolonged by low-dose insulin sensitization with metformin with beneficial and persisting effects on height and body composition [26]. Such pharmacological studies may also represent models for future behavioural intervention studies to reduce weight and insulin resistance during adolescence and thereby have possible benefits on avoiding early pubertal development and also potential long-term benefits on adult height, body composition, and metabolic disease risks.

Common Genetic Determinants of Childhood Obesity

The heritability of BMI increases during early childhood. A recent twins study showed that postnatal weight gain from birth to age 8 years was highly heritable (80%) and that there was a significant overlap between the heritability of postnatal weight gain, adiposity and fasting insulin [27]. A twin analyses on repeated assessments of BMI in a longitudinal sample of >7,000 children indicated that the genetic influence on BMI became progressively stronger, with heritability increasing from 0.48 at age 4 years to 0.78 at age 11 years [28], which is around the upper estimate of the heritability of BMI in adults [29].

Genome-wide association (GWA) studies have recently identified common genetic variants that are robustly associated with higher BMI in adults. A GWA study initially for type 2 diabetes identified a common variant in *FTO* (rs9939609) which conferred increased risk for diabetes secondary to its associations with greater adult BMI [30]. A second GWA study reported a common variant (rs17782313) 200 kb downstream of its nearest gene *MC4R* to give the next strongest association with adult BMI [31]. These GWA gene discovery studies for adult BMI/obesity have had major relevance for childhood BMI and body composition.

Through analysis of childhood data in the ALSPAC study it has been shown that common genetic variants near to *MC4R* had much stronger associations with BMI in children than in adults [31]. In the ALSPAC population-based cohort and in the SCOOP study of severely obese children, four out of the six novel adult BMI variants were also associated with BMI or obesity in children [32]. Furthermore, in those papers the childhood data showed that the effects of these genetic variants on body composition were largely on body fat rather than on fat-free mass [31, 32]. These findings are similar to those initially reported with *FTO* [30], and they support the notion that many of the specific genetic variants for adult obesity susceptibility have major effects on childhood weight gain.

Life-Course Associations with *FTO* and *MC4R* Variants

The timing of associations between common genetic variants for body weight or BMI across the life-course may provide insights into the aetiology of obesity. A unique study of a longitudinal birth cohort born in 1946 and followed through to age 53 years observed that the associations of *FTO* rs9939609 and *MC4R* rs17782313 on body size varied with age [33]. The changes with age appeared to be biphasic, in that the associations with both variants strengthened during childhood and adolescence, the associations peaked at age 20 years and thereafter weakened with increasing adult age (fig. 1) [33].

The authors of that study postulated that these apparent age-related changes may reflect the age-dependent influence of satiety and eating behaviour on weight gain and BMI, as weight gain in adult life may be increasingly determined or modified by other hedonic, psychosocial or environmental influences [34]. Studies of the aetiology of obesity spanning different age groups could therefore identify age-specific determinants of weight gain. However, further longitudinal population-based studies spanning a wide range of birth years are required to distinguish between the effects of age and period changes in the obesogenic environment.

Nutritional Regulation of Infancy Growth

Identification of the non-genetic determinants of infant weight may suggest potential future targets for intervention to prevent rapid infancy weight gain and childhood obesity. The ALSPAC study previously reported that breast-fed infants showed markedly lower weight gain from birth than formula-fed infants [35]. That study went on to show that within the formula-fed infants, total energy intake was positively associated with rate of weight gain and subsequent childhood BMI [36]. While such associations between weight gain and energy intake

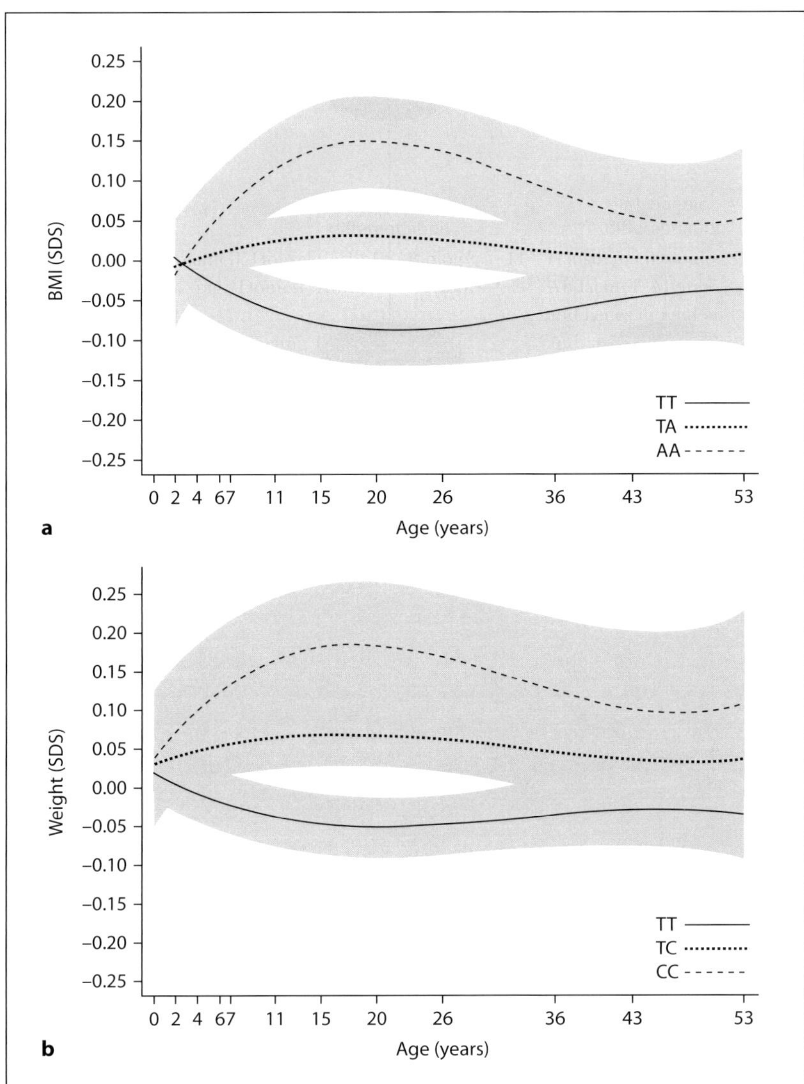

Fig. 1. Prediction of mean (±95% confidence interval) from additive genetic models for (**a**) BMI SDS by *FTO* rs9939609 genotype and (**b**) weight *SDS* by *MC4R* rs17782313 genotype, from birth to age 53 years in the National Survey of Health and Development [reprinted from 33].

might have been 'assumed', they are remarkably difficult to show at other ages in children or adults, likely due to the wide interindividual variation in energy requirements. Those 'positive' findings in infants support the greater dependency between weight gain and nutrition during infancy than at older ages [37], and suggest that changes in calorie intake in infants could alter both rate of weight gain and subsequent BMI.

Biomarkers of Infant Growth and Adiposity

A recent report from the Cambridge Baby Growth Study confirmed the effects of breast-feeding on slower weight gain, and also demonstrated that formula milk-fed infants grew faster in all aspects of infant growth, including length, BMI and skinfolds [38]. In order to identify potential non-nutritional biomarkers on infant growth and weight gain, IGF-1 levels were measured in 953 capillary dried blood spot samples from 675 infants from that study. IGF-1 is a major childhood growth factor and promotes bone and skeletal muscle growth. IGF-1 levels were higher in formula milk-fed infants and explained why these infants grew faster in length, however IGF-1 levels were apparently protective against gains in BMI and adiposity [38]. Therefore, higher infant IGF-I levels may indicate which infants will partition their weight gain more into lean mass than into fat mass. However, other factors independent of IGF-1 must explain the link between formula milk feeding and faster gains in adiposity.

Influence of Maternal Glucose on Birth Size and Infancy Growth

It has been suggested that maternal glucose metabolism during pregnancy might have long-term programming effects on obesity risk in her offspring. However, such observations may be confounded by prenatal factors (e.g. mother's BMI) and shared postnatal environment. In 668 mothers from the Cambridge Baby Growth Study, the influence of mothers' glucose and insulin levels during pregnancy was explored on the offspring's body weight and adiposity at birth and during infancy [39]. After adjustment for pre-pregnancy BMI, mother's higher fasting glycaemia was related to greater offspring adiposity at birth. However, during postnatal follow-up, the correlation between mother's glycaemia and offspring adiposity disappeared by 3 months, whereas mother's pre-pregnancy BMI was associated with offspring adiposity that was only apparent at 12 and 24 months (both p < 0.05). These studies suggest that influences associated with mother's BMI, such as genetic factors or shared environment, become increasingly apparent on offspring adiposity soon after birth. In contrast, the influence of maternal glycaemia on offspring adiposity, as assessed by skinfold measurements, was only transient [39].

Wider Determinants of Infant Feeding and Weight Gain

In order to inform the development of complex behavioural interventions, studies are needed to explore the potentially mutable determinants of infant feeding and weight gain. Such factors may include non-genetic individual factors, such as infant diet and nutrition as discussed above. Consideration should also be given to 'wider' determinants, such as government policies on infant measurement and feeding, and the reasons why such policies may or may not be followed by parents.

WHO Infant Growth Standard. In 2006 the WHO published the first international infancy growth charts using predominantly breast-fed infants as the standard for optimal growth and nutrition. By modelling data from the ALSPAC and Gateshead Millennium cohorts, it was demonstrated that the application of such charts in the UK would increase by 35% the proportion of infants who are classified as overweight and would substantially reduce by around 85% the proportion of infants currently (wrongly) classified as underweight [40]. These studies indicate the potential major relevance of use of appropriate growth references or standards to set a lower trajectory of infant weight gain and avoid future childhood obesity. Furthermore, promotion of breast-fed infants as the standard for growth supports the notion that formula milk-fed infants are being overfed. These studies contributed to the UK Department of Health Scientific Advisory Committee on Nutrition's report on the application of WHO growth standards in the UK and the subsequent development of the UK version of these charts.

Determinants of Non-Recommended Feeding Practices. Simply setting new recommendations for infant feeding is unlikely to be sufficient to change parental behaviours. A recent systematic review identified those factors associated with non-recommended infant feeding practices such as short duration of breast-feeding, early introduction of solid foods, and early use of unmodified cow's milk [41]. That review summarized data from 78 studies from developed settings, and found strong evidence for six determinants of early weaning, namely: young maternal age, low maternal education, low socioeconomic status, absence or short duration of breast-feeding, maternal smoking, and lack of information or advice from healthcare providers. Two determinants were also identified for early introduction of unmodified cow's milk: low maternal education and low socioeconomic status. Of these determinants, improving advice given by healthcare providers appeared the most tractable area for intervention in the short term. However, there was a paucity of studies exploring the psychosocial, community and policy factors on infant feeding practices that might inform the development of interventions at multiple levels [41].

Determinants of Parental Feeding Choices

A further systematic review summarized the quantitative and qualitative studies that specifically addressed how mothers who bottle-fed their infants made decisions regarding feeding [42]. Six qualitative studies and 17 quantitative studies (involving 13,263 participants) were included. Despite wide differences in study design, context, focus and quality, several consistent themes emerged. Mothers who bottle-fed their babies experienced negative emotions such as guilt, anger, worry, uncertainty and a sense of failure. Mothers reported receiving little information on bottle-feeding and did not feel empowered to make decisions. These studies will usefully inform the development of interventions to optimize feeding and weight gain in formula milk-fed infants.

References

1 Reilly JJ, Methven E, McDowell ZC, et al: Health consequences of obesity. Arch Dis Child 2003;88: 748–752.
2 Wardle J, Brodersen NH, Cole TJ, et al: Development of adiposity in adolescence: five-year longitudinal study of an ethnically and socioeconomically diverse sample of young people in Britain. BMJ 2006;332:1130–1135.
3 Baker JL, Olsen LW, Sorensen TI: Childhood body mass index and the risk of coronary heart disease in adulthood. N Engl J Med 2007;357: 2329–2337.
4 Summerbell CD, Waters E, Edmunds LD, et al: Interventions for preventing obesity in children. Cochrane Database Syst Rev CD001871 (2005).
5 Buchan IE, Bundred PE, Kitchiner DJ, Cole TJ: Body mass index has risen more steeply in tall than in short 3-year-olds: serial cross-sectional surveys 1988–2003. Int J Obes (Lond) 2007;31:23–29.
6 Ong KK, Ahmed ML, Emmett PM, et al: Association between postnatal catch-up growth and obesity in childhood: prospective cohort study. BMJ 2000;320:967–971.
7 Baird J, Fisher D, Lucas P, et al: Being big or growing fast: systematic review of size and growth in infancy and later obesity. BMJ 2005;331:929.
8 Monteiro PO, Victora CG: Rapid growth in infancy and childhood and obesity in later life – a systematic review. Obes Rev 2005;6:143–154.
9 Ong KK, Loos RJ: Rapid infancy weight gain and subsequent obesity: systematic reviews and hopeful suggestions. Acta Paediatr 2006;95:904–908.
10 Ekelund U, Ong K, Linné Y, et al: Upward weight percentile crossing in infancy and early childhood independently predicts fat mass in young adults: the Stockholm Weight Development Study (SWEDES). Am J Clin Nutr 2006;83:324–330.
11 Ekelund U, Ong KK, Linné Y, et al: Association of weight gain in infancy and early childhood with metabolic risk in young adults. J Clin Endocrinol Metab 2007;92:98–103.
12 Ibanez L, Ong K, Dunger DB, de Zegher F: Early development of adiposity and insulin resistance after catch-up weight gain in small-for-gestational-age children. J Clin Endocrinol Metab 2006;91:2153–2158.
13 Karaolis-Danckert N, Buyken AE, Bolzenius K, et al: Rapid growth among term children whose birth weight was appropriate for gestational age has a longer lasting effect on body fat percentage than on body mass index. Am J Clin Nutr 2006;84: 1449–1455.
14 Mericq V, Ong KK, Bazaes R, et al: Longitudinal changes in insulin sensitivity and secretion from birth to age three years in small- and appropriate-for-gestational-age children. Diabetologia 2005; 48:2609–2614.
15 Honour JW, Jones R, Leary S, et al: Relationships of urinary adrenal steroids at age 8 years with birth weight, postnatal growth, blood pressure, and glucose metabolism. J Clin Endocrinol Metab 2007;92:4340–4345.
16 Ong KK, Elmlinger M, Jones R, et al: Growth hormone binding protein levels in children are associated with birth weight, postnatal weight gain, and insulin secretion. Metabolism 2007;56:1412–1417.
17 Ong KK, Frystyk J, Flyvbjerg A, et al: Sex-discordant associations with adiponectin levels and lipid profiles in children. Diabetes 2006;55:1337–1341.
18 Singhal A, Cole TJ, Fewtrell M, et al: Is slower early growth beneficial for long-term cardiovascular health? Circulation 2004;109:1108–1113.
19 Cross-Government Obesity Unit, DoHa DoC, Schools and Families. Healthy Weight, Healthy Lives: a Cross-Government Strategy for England. London, Department of Health/HMSO, 2008.
20 Ong KK, Emmett P, Northstone K, et al: Infancy weight gain predicts childhood body fat and age at menarche in girls. J Clin Endocrinol Metab 2009;94:1527–1532.

21 Ong KK, Northstone K, Wells JC, et al: Earlier mother's age at menarche predicts rapid infancy growth and childhood obesity. PLoS Med 2007;4: e132.

22 Lakshman R, et al: Association between age at menarche and risk of diabetes in adults: results from the EPIC-Norfolk cohort study. Diabetologia 2008;51:781–786.

23 Lakshman R, Forouhi N, Luben R, et al: Early age at menarche associated with cardiovascular disease and mortality. J Clin Endocrinol Metab 2009; 94:4953–4960.

24 Ibanez L, Ong K, de Zegher F, et al: Fat distribution in non-obese girls with and without precocious pubarche: central adiposity related to insulinaemia and androgenaemia from prepuberty to postmenarche. Clin Endocrinol (Oxford) 2003;58:372–379.

25 Ibanez L, Ong K, Valls C, et al: Metformin treatment to prevent early puberty in girls with precocious pubarche. J Clin Endocrinol Metab 2006;91: 2888–2891.

26 Ong K, de Zegher F, Valls C, et al: Persisting benefits 12–18 months after discontinuation of pubertal metformin therapy in low birthweight girls. Clin Endocrinol (Oxf) 2007;67:468–471.

27 Beardsall K, Ong KK, Murphy N, et al: Heritability of childhood weight gain from birth and risk markers for adult metabolic disease in prepubertal twins. J Clin Endocrinol Metab 2009;94:3708–3713.

28 Haworth CM, Carnell S, Meaburn EL, et al: Increasing heritability of BMI and stronger associations with the FTO gene over childhood. Obesity (Silver Spring) 2008;16:2663–2668.

29 Maes HH, Neale MC, Eaves LJ: Genetic and environmental factors in relative body weight and human adiposity. Behav Genet 1997;27:325–351.

30 Frayling TM, Timpson NJ, Weedon MN, et al: A common variant in the FTO gene is associated with body mass index and predisposes to childhood and adult obesity. Science 2007;316:889–894.

31 Loos RJ, Lindgren CM, Li S, et al: Common variants near MC4R are associated with fat mass, weight and risk of obesity. Nat Genet 2008;40:768–775.

32 Willer CJ, Speliotes EK, Loos RJ, et al: Six new loci associated with body mass index highlight a neuronal influence on body weight regulation. Nat Genet 2009;41:25–34.

33 Hardy R, Wills AK, Wong A, et al: Life course variations in the associations between FTO and MC4R gene variants and body size. Hum Mol Genet 2010;19:545–552.

34 O'Rahilly S, Farooqi IS: Human obesity: a heritable neurobehavioral disorder that is highly sensitive to environmental conditions. Diabetes 2008; 57:2905–2910.

35 Ong KK, Preece MA, Emmett PM, et al: Size at birth and early childhood growth in relation to maternal smoking, parity and infant breast-feeding: longitudinal birth cohort study and analysis. Pediatr Res 2002;52:863–867.

36 Ong KK, Emmett PM, Noble S, et al: Dietary energy intake at the age of 4 months predicts postnatal weight gain and childhood body mass index. Pediatrics 2006;117:e503–508.

37 FAO: Human Energy Requirements: Report of a Joint FAO/WHO/UNU Expert Consultation. Food Nutr Tech Rep Ser 1, 2004.

38 Ong K, Langkamp M, Ranke MB, et al: Insulin-like growth factor-I concentrations in infancy predict differential gains in body length and adiposity: the Cambridge Baby Growth Study. Am J Clin Nutr 2009;90:156–161.

39 Ong KK, Diderholm B, Salzano G, et al: Pregnancy insulin, glucose, and BMI contribute to birth outcomes in nondiabetic mothers. Diabetes Care 2008;31:2193–2197.

40 Wright C, Lakshman R, Emmett P, Ong K: Implications of adopting the WHO 2006 Child Growth Standard in the UK: two prospective cohort studies. Arch Dis Child 2007;93:566–569.

41 Wijndaele K, Lakshmann R, Landsbaugh J, et al: Determinants of early weaning and use of unmodified cow's milk in infants: a systematic review. J Am Diet Assoc in press 2009;109:2017–2028.

42 Lakshman R, Ogilvie D, Ong K: Mothers' experiences of bottle-feeding: a systematic review of qualitative and quantitative studies. Arch Dis Child 2009;94:4953–4960.

Dr. Ken Ong
MRC Epidemiology Unit, Institute of Metabolic Science
Addenbrooke's Hospital, Box 285, Cambridge CB2 0QQ (UK)
Tel. +44 1223 769207, Fax +44 1223 330316
E-Mail ken.ong@mrc-epid.cam.ac.uk

Levy-Marchal C, Pénicaud L (eds): Adipose Tissue Development: From Animal Models to Clinical Conditions.
Endocr Dev. Basel, Karger, 2010, vol 19, pp 62–72

Metabolic Syndrome in Childhood – Causes and Effects

Ram Weiss

Hebrew University-Hadassah Braun School of Public Health and Community Medicine, Faculty of Medicine,
Jerusalem, Israel

Abstract

The metabolic syndrome defines the clustering of cardiovascular risk factors and is a driven by
peripheral insulin resistance. The 'driving force' of the syndrome, i.e. insulin resistance, develops
mainly in obese children due to a specific pattern of lipid partitioning characterized by increased
deposition of fat in the visceral compartment as well as in insulin-responsive tissues, such as
muscle and liver. Such a lipid deposition pattern results in peripheral insulin resistance and a
compensatory hyperinsulinemia. Hyperinsulinemia results in normal response of tissues and
metabolic pathways that maintained their insulin sensitivity, leading to the typical biochemical
and clinical manifestations of the metabolic syndrome. The definition of the syndrome in childhood suffers from many limitations related to different ethnic characteristics as well as age and
development dependency of some of the components. Despite these limitations, the clustering
of risk factors characteristic of the syndrome in childhood is associated with accelerated atherogenesis in adulthood. Thus, using threshold-based definitions of the syndrome is still important,
useful and practical for the sake of risk stratification, longitudinal follow-up and potentially for
treatment considerations.

The metabolic syndrome, also known as 'the insulin resistance' syndrome, describes
a cluster of cardiovascular risk factors that have been shown to predict the development of cardiovascular disease (CVD) [1, 2] and type 2 diabetes (T2DM) [3].
As depicted by its names, this syndrome is characterized by the consequences of
peripheral insulin resistance [4]. These metabolic consequences represent a normal physiological adaptation induced by the resistance of specific tissues to the
peripheral action of insulin. The clinical utility of defining this syndrome in adults
and in children has been debated, as some propose that from a clinical stand-
point, using each component of the syndrome individually has comparable clinical and predictive outcomes. This debate has important clinical implications yet

it is imperative to indicate that the underlying physiology that leads to the typical metabolic milieu characteristic of individuals with peripheral insulin resistance has common features in all ages and is postulated to be the 'driving force' of the development of accelerated atherogenesis and altered glucose metabolism in susceptible individuals. While insulin resistance can be induced by different metabolic stimuli during the life course (transient peripheral insulin resistance induced during pubertal development in adolescents [5], pregnancy in females of childbearing age, the effects of ageing per se in older individuals and consumption of specific dietary components [6]), the tissue-specific metabolic adaptation to such resistance has common mechanistic features regardless of age. The clinical manifestations of such an adaptation may rely on the length of exposure to it and on the susceptibility of the individual, yet the biomarkers of its presence exist at all ages.

Pathophysiology of the Insulin Resistance Syndrome in Childhood

Several groups have performed a factor analysis of components of the metabolic syndrome in cohorts of adults [7–9] and of children [10–12] in order to shed light on the typical associations observed between its components. These studies revealed that obesity or its related peripheral insulin resistance seems to cluster with the majority of the clinically used components of the syndrome yet also clusters with additional elements, such as increased fibrinolysis [13], endothelial dysfunction and subclinical inflammation [14] that seem to be part of the typical metabolic milieu of the insulin-resistant individual yet are not routinely assessed or clinically utilized for treatment decisions or for risk stratification. The conclusion of the majority of these analyses performed in children is that obesity, tightly associated with insulin resistance, is the common feature that is associated with the presence of the routinely measured cardiovascular risk factors.

It is important to indicate though that the presence of obesity or even severe obesity in a child does not necessarily indicate that significant insulin resistance is present and therefore does not imply that clustering of cardiovascular risk factors necessarily exists. Indeed, some severely obese children can be extremely insulin-sensitive [15]. On the other hand, some may possess other cardiovascular risk factors not necessarily associated with obesity or insulin resistance such as certain heritable dyslipidemias [16]. The relation of obesity and peripheral insulin resistance is dependent more on the lipid distribution (or 'lipid partitioning') in specific fat depots rather than on the absolute amount of fat per se. Different lipid depots have distinct metabolic characteristics that are reflected by their adipocytokine and cytokine secretion profile [17], sensitivity to hormones typically affecting adipose tissue (such as norepinephrine or insulin) as well as anatomical blood supply and drainage (portal vs. systemic) [18]. Indeed, increased visceral fat accumulation

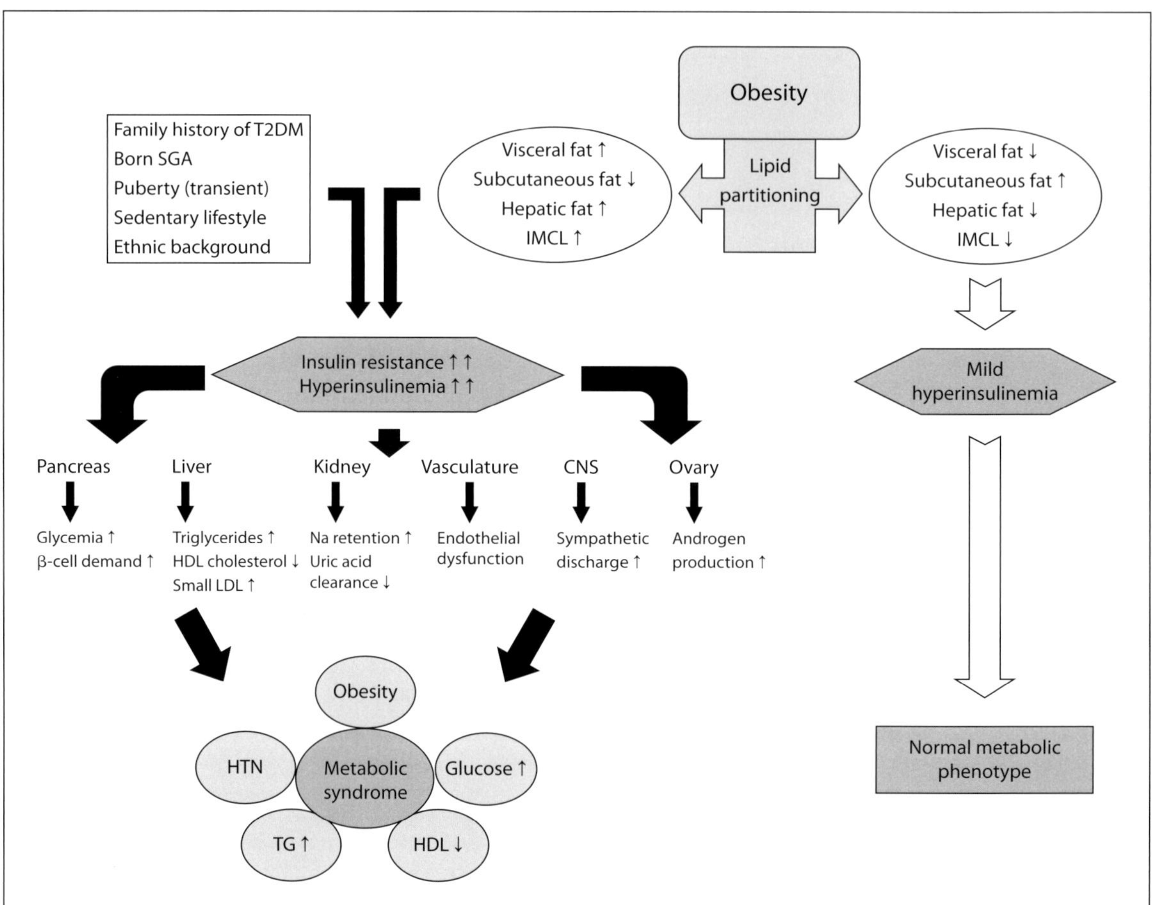

Fig. 1. Pathophysiology of the metabolic syndrome in childhood. The major cause of insulin resistance in childhood is obesity. The pattern of lipid partitioning along with genetically determined and environmental factors associated with insulin resistance (such as a family history of T2DM, being born small for gestational age (SGA), specific ethnic backgrounds as well as the pubertal period, determines the degree of peripheral insulin resistance. Those with severe insulin resistance respond physiologically with a compensatory hyperinsulinemia. This results in a 'normal' response of tissues and metabolic pathways that remain insulin-sensitive, manifesting as a variety of metabolic responses that lead to the development of the typical phenotype of insulin resistance – the metabolic syndrome. Those obese individuals with a favorable lipid portioning pattern and less genetic and environmental risk factors develop a mild insulin resistance and remain with a normal metabolic phenotype.

in obese children has been associated with increased insulin resistance [19] and with the clustering of cardiovascular risk factors as well as with worsening of each one of them individually [20]. Some obese children tend to demonstrate a unique lipid partitioning pattern characterized by a large visceral fat depot along with a relatively smaller subcutaneous fat depot. This lipid partitioning profile is associated with an adverse metabolic profile in comparison to those with larger subcutaneous fat depots, even when the latter have greater body mass index (BMI) and percent body fat and may thus be seemingly 'more obese' [21].

Lipid deposition in insulin-sensitive tissues, such as muscle and liver, represents another determinant of the sensitivity of these tissues to the metabolic effects of insulin. Intramyocellular lipid deposition is well correlated with peripheral insulin sensitivity and has been demonstrated to be increased in offspring of patients with T2DM and in obese children with impaired glucose tolerance [22]. The effect of lipid within the myocyte on insulin signal transduction is indirect and probably mediated by several fatty acid derivatives such as fatty acyl CoA that inhibit insulin signaling [23]. Similarly, hepatic fat accumulation is strongly associated with obesity and with hepatic resistance to the action of insulin in the context of pathways related to glucose metabolism and is also associated with an adverse cardiovascular risk profile in children. As the liver governs glucose metabolism in the fasting state and muscle governs glucose metabolism in the postabsorptive state, significant hepatic insulin resistance may have a stronger impact on fasting glucose levels and on the early postabsorptive suppression of hepatic glycogenolysis/gluconeogenesis following a meal, while muscle lipid deposition may have a greater impact on postprandial glucose levels. As both tissues develop insulin resistance in association with increased lipid deposition, the normal adaptive response consists of increased insulin secretion by the β cells along with reduced insulin clearance by the liver. The result of these two normal adaptive responses is increased circulating insulin levels (hyperinsulinemia). The ensuing hyperinsulinemia, at least at early stages of this metabolic adaptation, overcomes the tissue-specific insulin resistance and represents an adequate allostatic response, yet carries a 'metabolic price' of increased circulating glucose levels along with a continuous burden on the β cells to secrete adequate amounts of insulin [24, 25]. Importantly, as shown in figure 1, other metabolic pathways within the liver that are not involved in glucose metabolism or other insulin-sensitive tissues that do not share the pattern of increased lipid deposition within them, such as the kidney or the ovary, maintain their baseline insulin sensitivity levels yet are now exposed to hyperinsulinemia. This exposure may result in a normal response of these tissues to elevated insulin levels and manifest as increased sodium retention and reduced uric acid clearance by the kidney [26] and by increased androgen production by the ovary [27]. Insulin may also induce an activation of the sympathetic nervous system [28] and impact the metabolism and secretion of pro-inflammatory cytokines as well as coagulation mediators [29]. Similarly, other metabolic pathways within the liver, specifically those related to lipoprotein metabolism, may maintain their baseline insulin sensitivity (unlike those pathways related to glucose metabolism) and thus respond to the elevated insulin levels in a pattern that creates the typical dyslipidemia characteristic of insulin-resistant individuals. Some suggest that hepatic deposition of lipid is not a primary process but a 'normal' response to elevated circulating insulin levels induced by muscle insulin resistance and thus that hepatic steatosis is part of the result and not one of the culprits of the adverse metabolic phenotype characteristic of insulin-resistant individuals.

In summary, peripheral tissue resistance to the action of insulin, specifically in metabolic pathways related to glucose metabolism, results in a compensatory hyperinsulinemia. The exposure of metabolic pathways and tissues that were not affected by obesity or local lipid deposition and maintained their original sensitivity to insulin results in a metabolic profile characterized by the presence of dyslipidemia (specifically elevated triglycerides, low HDL cholesterol and the presence of small LDL particles), elevated blood pressure and altered glucose metabolism (that may manifest as impaired fasting glucose, impaired glucose tolerance or overt diabetes). In addition, pro-inflammatory cytokines and factors related to hypercoagulability may be present at increased levels. The clustering of some or all of these factors results in accelerated atherogenesis leading to later CVD and in continuous β-cell stress that in some individuals may result in β-cell failure and hyperglycemia (an individual contributor to accelerated atherogensis).

The Complexity of Definitions of the Metabolic Syndrome

The metabolic syndrome in children has several definitions used by various research groups [30, 31] and the use of different definitions in the same cohort may result in different outcomes [32]. All of these definitions share several common features: these include a component representing obesity (such as waist circumference or BMI), the typical dyslipidemia characteristic of the syndrome (elevated triglycerides and low HDL cholesterol), elevated blood pressure and an element related to glucose metabolism (such as impaired fasting glucose). The recently published consensus definition [33] utilizes waist circumference, a surrogate of abdominal obesity, as the body habitus component, thus reflecting the paramount importance of intra-abdominal fat for the development of specific cardiovascular risk factors. One must question the usefulness of such definitions in their context of utilization, i.e. in the clinical setting or for research purposes. Some have questioned the clinical utility of such definitions in adults [34] and in children [35] and advocate addressing individual risk factors in their clinical context. Regardless of this controversy, one must appreciate that cardiovascular risk factors such as elevated fasting glucose or the degree of obesity represent continuous variables that signify risk, not necessarily in a linear fashion. Thus, for example, while increasing BMI during childhood represents a continuous risk factor for the development of coronary heart disease in adulthood even within the normal BMI range [36], severely obese children may have a significantly worse metabolic phenotype in comparison to moderately obese children [37]. On the other hand, a seemingly 'upper normal' fasting glucose in the context of obesity may signify future risk [38]. Similarly, the triglyceride level in late adolescence and its change within a brief follow-up of ~5 years can predict the development of diabetes [39] and of coronary heart disease [40] even when both measurements are within the seemingly 'normal' range.

Another issue that adds complexity to definitions of the metabolic syndrome is the problematic generalizability for populations of a different ethnic background. In parallel to the rise of the prevalence of obesity, the emergence of T2DM in children and adolescents is more common in ethnic minorities in the USA [41] as well as in some European countries [42]. A potential explanation for the ethnic disparities in the incidence of obesity-related diseases such as T2DM in adolescence is that youth of ethnic minorities are more obese and more insulin-resistant in comparison to their Caucasian peers [43]. Differences in insulin secretion and clearance with similar peripheral sensitivity have been demonstrated in non-obese Hispanic- and African-Americans in comparison to Caucasian children [44]. These differences translate into higher circulating insulin levels in such children (i.e. 'relative hyperinsuline-mia') and may thus promote the normal response of non-insulin-resistant tissues to elevated insulin concentrations, as depicted earlier. This implies that thresholds for surrogates of insulin sensitivity to be used in definitions of the metabolic syn-drome that utilize a fasting or postprandial insulin level should be ethnicity-spe-cific. Similarly, individuals with a comparable BMI of different ethnic background may have a significantly different body composition. Such differences translate into a greater percent body fat per given BMI leading to an increased vulnerability to the adverse impact of increased specific body fat depots in individuals at lower BMI or waist circumference levels [45]. In addition, youth of different ethnic background may differ in their patterns of lipid partitioning in insulin-sensitive tissues such as muscle and liver [46] and thus at similar degrees of obesity may be entirely different with regard to their degree of insulin sensitivity. This implies that anthropometric measures to be used in definitions of the metabolic syndrome should be ethnicity-sensitive and derived from outcome data of the relevant population [47].

The present definitions of the metabolic syndrome in children and in adults use components with equal value within the scoring system that add up to a sum of risk factors that provide the 'metabolic syndrome score' which should pass a cumulative threshold in order to meet the criteria of the diagnosis. This definition structure does not take into account that individual components of the definition may carry a differ-ent effect size on future risk nor that individuals of different ethnic background may display unique components of the syndrome in relation to the sensitivity of specific metabolic pathways to the impact of hyperinsulinemia. De Ferranti et al. [31] found that low HDL and elevated triglycerides are the most common metabolic abnor-malities present in adolescents from a population-based cohort, yet the prevalence of these two elements of insulin resistance-related dyslipidemia seems to differ between youth of different ethnic backgrounds [48]. The higher triglyceride levels observed in Caucasian youth seem to result from differences in VLDL subparticles, specifically greater large VLDL particles [49] while African-Americans have been shown to have larger LDL particles [50]. In contrast to their seemingly favorable lipid profile, African-American children and adolescents have a greater systolic blood pressure per given degree of obesity [51] and display differences in renal handling of potassium in comparison to their Caucasian peers. These observations imply that youth of different ethnic background have significant differences in body composition, lipid partitioning patterns and sensitivity of different tissues or metabolic pathways to similar circulating insulin levels. This adds further complexity to the utilization of threshold-based definitions of the metabolic syndrome in childhood and emphasizes the difficulty of creating a 'one fits all' definition for clinical practice.

Despite the fact that utilization of definitions that are based on thresholds of specific risk elements may undermine the characteristic continuum of these risk factors and result in under- or overdiagnosis when used in populations of different ethnic background, it still provides a simple tool of risk stratification for the clinical setting. The caregiver using such definitions is prompted to understand the common pathophysiology underlying their clustering and to search for elements of such clustering that do not have an obvious clinical presentation (such as dyslipidemia). Thus, despite numerous limitations, some of which have been presented here, definitions of the syndrome in children and adolescents have a major importance in the clinical as well as the research setting. Utilization of such definitions can facilitate the identification of those obese children that are most vulnerable and carry the greatest risk of the adverse impact of obesity in childhood and who may benefit most from effective interventions. As such interventions are costly and intensive, identification of those who may benefit most from them may facilitate their success.

Clinical Relevance

Children who meet criteria of the metabolic syndrome have been shown to have increased aortic stiffness [52] while associations with endothelial dysfunction (a surrogate of an early accelerated atherogenic process) show mixed results [53, 54]. Regardless of the definitions used, the presence of cardiovascular risk factors in childhood and adolescence tracks into adulthood and is associated with the presence of subclinical atherosclerosis in young adulthood [55, 56]. Although the stability of the diagnosis of the metabolic syndrome in none obese children is questionable [57], the stability of the syndrome in obese children (i.e. those at greatest risk to develop it) is tightly linked to dynamics in the degree of obesity and insulin sensitivity [58]. Thus, in those at greatest risk, definitions of the metabolic syndrome are useful for diagnostic as well as for follow-up evaluation.

These observations highlight the importance of primary and secondary prevention of the progression of early cardiovascular risk factors in children and adolescents. As the presence of some of the components of the metabolic syndrome tends to track from childhood to adulthood, primary prevention of their development or early reversal of their presence in childhood are of paramount importance. Interventions ranging from diet-induced weight loss and increased physical activity to bariatric surgery have been attempted in obese children with the presence of cardiovascular risk factors or overt disease such as T2DM. Such interventions have shown that the level of cardiovascular risk factors related to the metabolic syndrome can be reduced [59] and that presence of T2DM can be eliminated [60]. As these interventions are expensive and labor-intensive, selection of those obese youth who may benefit most from them is crucial. Measures such as risk factor clustering (metabolic syndrome definitions) and its dynamics over time can serve as selection and follow-up tools for such patients.

Conclusion

Clustering of cardiovascular risk factors, the development of which is driven by peripheral insulin resistance, is present in children and adults. Such clustering predicts the presence of adverse outcomes in adulthood. Obesity per se in a child does not necessarily mean that insulin resistance is present. The pattern of lipid partitioning, adipocytokine profile and presence of genetically determined factors (such as ethnicity, family history of T2DM and others) is crucial for the development of the adverse metabolic phenotype typical of insulin-resistant children. Definitions of the syndrome that are based on thresholds may be useful in clinical practice for the identification and follow-up of those youth who may benefit most from therapeutic interventions. Yet, the elements related to the syndrome represent a continuum of risk and should thus be addressed and followed even when they are seemingly 'normal'.

References

1   Gami AS, Witt BJ, Howard DE, Erwin PJ, Gami LA, Somers VK, Montori VM: Metabolic syndrome and risk of incident cardiovascular events and death: a systematic review and meta-analysis of longitudinal studies. J Am Coll Cardiol 2007;49:403–414.

2   Pyorala M, Miettinen H, Halonen P, Laakso M, Pyorala K: Insulin resistance syndrome predicts the risk of coronary heart disease and stroke in healthy middle-aged men: the 22-year follow-up results of the Helsinki Policemen Study. Arterioscler Thromb Vasc Biol 2000;20:538–544.

3   Cornier MA, Dabelea D, Hernandez TL, Lindstrom RC, Steig AJ, Stob NR, Van Pelt RE, Wang H, Eckel RH: The metabolic syndrome. Endocr Rev 2008;29:777–822.

4   Reaven GM: Role of insulin resistance in human disease. Diabetes 1988;37:1595–1607.

5   Goran MI, Gower BA: Longitudinal study on pubertal insulin resistance. Diabetes 2001;50:2444–2450.

6   Lutsey PL, Steffen LM, Stevens J: Dietary intake and the development of the metabolic syndrome: the Atherosclerosis Risk in Communities Study. Circulation 2008;117:754–761.

7   Sakkinen PA, Wahl P, Cushman M, Lewis MR, Tracy RP: Clustering of procoagulation, inflammation, and fibrinolysis variables with metabolic factors in insulin resistance syndrome. Am J Epidemiol 2000;152:897–907.

8   Hanley AJ, Festa A, D'Agostino RB Jr, Wagenknecht LE, Savage PJ, Tracy RP, Saad MF, Haffner SM: Metabolic and inflammation variable clusters and prediction of type 2 diabetes: factor analysis using directly measured insulin sensitivity. Diabetes 2004;53:1773–1781.

9   Meigs JB, D'Agostino RB Sr, Wilson PW, Cupples LA, Nathan DM, Singer DE: Risk variable clustering in the insulin resistance syndrome: the Framingham Offspring Study. Diabetes 1997;46:1594–600.

10 Li C, Ford ES: Is there a single underlying factor for the metabolic syndrome in adolescents? A confirmatory factor analysis. Diabetes Care 2007;30:1556–1561.

11 Goodman E, Dolan LM, Morrison JA, Daniels SR: Factor analysis of clustered cardiovascular risks in adolescence: obesity is the predominant correlate of risk among youth. Circulation 2005;111:1970–1977.

12 Weiss R, Dziura J, Burgert TS, Tamborlane WV, Taksali SE, Yeckel CW, Allen K, Lopes M, Savoye M, Morrison J, Sherwin RS, Caprio S: Obesity and the metabolic syndrome in children and adolescents. N Engl J Med 2004;350:2362–2374.

13 Mertens I, Van Gaal LF: Obesity, haemostasis and the fibrinolytic system. Obes Rev 2002;3:85–101.

14 Yudkin JS: Inflammation, obesity, and the metabolic syndrome. Horm Metab Res 2007;39:707–709.

15 Weiss R, Taksali SE, Dufour S, Yeckel CW, Papademetris X, Cline G, Tamborlane WV, Dziura J, Shulman GI, Caprio S: The 'obese insulin-sensitive' adolescent: importance of adiponectin and lipid partitioning. J Clin Endocrinol Metab 2005;90:3731–3737.

16 Kwiterovich PO Jr: Recognition and management of dyslipidemia in children and adolescents. J Clin Endocrinol Metab 2008;93:4200–4209.

17 Matsuzawa Y, Funahashi T, Nakamura T: Molecular mechanism of metabolic syndrome X: contribution of adipocytokines adipocyte-derived bioactive substances. Ann NY Acad Sci 2000;893:146–154.

18 Wajchenberg BL: Subcutaneous and visceral adipose tissue: their relation to the metabolic syndrome. Endocr Rev 2000;21:697–738.

19 Cruz ML, Bergman RN, Goran MI: Unique effect of visceral fat on insulin sensitivity in obese Hispanic children with a family history of type 2 diabetes. Diabetes Care 2002;25:1631–1636.

20 Bacha F, Saad R, Gungor N, Janosky J, Arslanian SA: Obesity, regional fat distribution, and syndrome X in obese black versus white adolescents: race differential in diabetogenic and atherogenic risk factors. J Clin Endocrinol Metab 2003;88:2534–2540.

21 Taksali SE, Caprio S, Dziura J, Dufour S, Calí AM, Goodman TR, Papademetris X, Burgert TS, Pierpont BM, Savoye M, Shaw M, Seyal AA, Weiss R: High visceral and low abdominal subcutaneous fat stores in the obese adolescent: a determinant of an adverse metabolic phenotype. Diabetes 2008;57:367–371.

22 Weiss R, Dufour S, Taksali SE, Tamborlane WV, Petersen KF, Bonadonna RC, Boselli L, Barbetta G, Allen K, Rife F, Savoye M, Dziura J, Sherwin R, Shulman GI, Caprio S: Prediabetes in obese youth: a syndrome of impaired glucose tolerance, severe insulin resistance, and altered myocellular and abdominal fat partitioning. Lancet 2003;362:951–957.

23 Shulman GI: Cellular mechanisms of insulin resistance. J Clin Invest 2000;106:171–176.

24 Stumvoll M, Tataranni PA, Stefan N, Vozarova B, Bogardus C: Glucose allostasis. Diabetes 2003;52:903–909.

25 Weiss R, Cali AM, Dziura J, Burgert TS, Tamborlane WV, Caprio S: Degree of obesity and glucose allostasis are major effectors of glucose tolerance dynamics in obese youth. Diabetes Care 2007;30:1845–1850.

26 Facchini F, Chen YD, Hollenbeck CB, Reaven GM: Relationship between resistance to insulin-mediated glucose uptake, urinary uric acid clearance, and plasma uric acid concentration. JAMA 1991;266:3008–3011.

27 Dunaif A: Insulin resistance and the polycystic ovary syndrome: mechanism and implications for pathogenesis. Endocr Rev 1997;18:774–800.

28 Anderson EA, Hoffman RP, Balon TW, Sinkey CA, Mark AL: Hyperinsulinemia produces both sympathetic neural activation and vasodilation in normal humans. J Clin Invest 1991;87:2246–2252.

29 Van Gaal LF, Mertens IL, De Block CE: Mechanisms linking obesity with cardiovascular disease. Nature 2006;444:875–880.

30 Druet C, Dabbas M, Baltakse V, Payen C, Jouret B, Baud C, Chevenne D, Ricour C, Tauber M, Polak M, Alberti C, Levy-Marchal C: Insulin resistance and the metabolic syndrome in obese French children. Clin Endocrinol 2006;64:672–678.

31 De Ferranti SD, Gauvreau K, Ludwig DS, Neufeld EJ, Newburger JW, Rifai N: Prevalence of the metabolic syndrome in American adolescents: findings from the Third National Health and Nutrition Examination Survey. Circulation 2004;110:2494–2497.

32 Lee S, Bacha F, Gungor N, Arslanian S: Comparison of different definitions of pediatric metabolic syndrome: relation to abdominal adiposity, insulin resistance, adiponectin, and inflammatory biomarkers. J Pediatr 2008;152:177–184.

33 Zimmet P, Alberti G, Kaufman F, Tajima N, Silink M, Arslanian S, Wong G, Bennett P, Shaw J, Caprio S: The metabolic syndrome in children and adolescents. International Diabetes Federation Task Force on Epidemiology and Prevention of Diabetes. Lancet 2007;369:2059–2061.

34 Kahn R, Buse J, Ferrannini E, Stern M; American Diabetes Association; European Association for the Study of Diabetes: The metabolic syndrome: time for a critical appraisal: joint statement from the American Diabetes Association and the European Association for the Study of Diabetes. Diabetes Care 2005;28:2289–2304.

35 Brambilla P, Lissau I, Flodmark CE, Moreno LA, Widhalm K, Wabitsch M, Pietrobelli A: Metabolic risk-factor clustering estimation in children: to draw a line across pediatric metabolic syndrome. Int J Obes (Lond) 2007;31:591–600.

36 Baker JL, Olsen LW, Sørensen TI: Childhood body mass index and the risk of coronary heart disease in adulthood. N Engl J Med 2007;357:2329–2337.

37 Freedman DS, Mei Z, Srinivasan SR, Berenson GS, Dietz WH: Cardiovascular risk factors and excess adiposity among overweight children and adolescents: the Bogalusa Heart Study. J Pediatr 2007;150:12–17.

38 Tirosh A, Shai I, Tekes-Manova D, Israeli E, Pereg D, Shochat T, Kochba I, Rudich A, Israeli Diabetes Research Group: Normal fasting plasma glucose levels and type 2 diabetes in young men. N Engl J Med 2005;353:1454–1462.

39 Tirosh A, Shai I, Bitzur R, Kochba I, Tekes-Manova D, Israeli E, Shochat T, Rudich A: Changes in triglyceride levels over time and risk of type 2 diabetes in young men. Diabetes Care 2008;31:2032–2037.

40 Tirosh A, Rudich A, Shochat T, Tekes-Manova D, Israeli E, Henkin Y, Kochba I, Shai I: Changes in triglyceride levels and risk for coronary heart disease in young men. Ann Intern Med 2007;147:377–385.

41 Dabelea D, Pettitt DJ, Jones KL, Arslanian SA: T2DM mellitus in minority children and adolescents: an emerging problem. Endocrinol Metab Clin North Am 1999;28:709–729.

42 Haines L, Wan KC, Lynn R, Barrett TG, Shield JP: Rising incidence of type 2 diabetes in children in the UK. Diabetes Care 2007;30:1097–1101.

43 Arslanian SA: Metabolic differences between Caucasian and African-American children and the relationship to T2DM mellitus. J Pediatr Endocrinol Metab 2002;15(suppl 1):509–517.

44 Arslanian SA, Saad R, Lewy V, Danadian K, Janosky J: Hyperinsulinemia in African-American children: decreased insulin clearance and increased insulin secretion and its relationship to insulin sensitivity. Diabetes 2002;51:3014–3019.

45 Deurenberg-Yap M, Chew SK, Deurenberg P: Elevated body fat percentage and cardiovascular risks at low body mass index levels among Singaporean Chinese, Malays and Indians. Obes Rev 2002;3:209–215.

46 Liska D, Dufour S, Zern TL, Taksali S, Calí AM, Dziura J, Shulman GI, Pierpont BM, Caprio S: Interethnic differences in muscle, liver and abdominal fat partitioning in obese adolescents. PLoS ONE 2007;2:e569.

47 Misra A, Wasir JS, Vikram NK: Waist circumference criteria for the diagnosis of abdominal obesity are not applicable uniformly to all populations and ethnic groups. Nutrition 2005;21:969–976.

48 Srinivasan SR, Frerichs RR, Webber LS, Berenson GS: Serum lipoprotein profile in children from a biracial community: the Bogalusa Heart Study. Circulation 1976;54:309–318.

49 Freedman DS, Bowman BA, Otvos JD, Srinivasan SR, Berenson GS: Differences in the relation of obesity to serum triacylglycerol and VLDL subclass concentrations between black and white children: the Bogalusa Heart Study. Am J Clin Nutr 2002;75:827–833.

50 Freedman DS, Bowman BA, Otvos JD, Srinivasan SR, Berenson GS: Levels and correlates of LDL and VLDL particle sizes among children: the Bogalusa Heart Study. Atherosclerosis 2000;152: 441–449.

51 Jago R, Harrell JS, McMurray RG, Edelstein S, El Ghormli L, Bassin S: Prevalence of abnormal lipid and blood pressure values among an ethnically diverse population of eighth-grade adolescents and screening implications. Pediatrics 2006;117: 2065–2073.

52 Iannuzzi A, Licenziati MR, Acampora C, Renis M, Agrusta M, Romano L, Valerio G, Panico S, Trevisan M: Carotid artery stiffness in obese children with the metabolic syndrome. Am J Cardiol 2006;97:528–531.

53 Lee S, Gungor N, Bacha F, Arslanian S: Insulin resistance: link to the components of the metabolic syndrome and biomarkers of endothelial dysfunction in youth. Diabetes Care 2007;30: 2091–2097.

54 Mimoun E, Aggoun Y, Pousset M, Dubern B, Bouglé D, Girardet JP, Basdevant A, Bonnet D, Tounian PJ: Association of arterial stiffness and endothelial dysfunction with metabolic syndrome in obese children. Pediatrics 2008;153:65–70.

55 Berenson GS, Srinivasan SR, Bao W, Newman WP III, Tracy RE, Wattigney WA: Association between multiple cardiovascular risk factors and atherosclerosis in children and young adults: the Bogalusa Heart Study. N Engl J Med 1998;338: 1650–1656

56 Mahoney LT, Burns TL, Stanford W, Thompson BH, Witt JD, Rost CA, Lauer RM: Coronary risk factors measured in childhood and young adult life are associated with coronary artery calcification in young adults: the Muscatine Study. J Am Coll Cardiol 1996;27:277–284

57 Goodman E, Daniels SR, Meigs JB, Dolan LM: Instability in the diagnosis of metabolic syndrome in adolescents. Circulation 2007;115:2316–2322.

58 Weiss R, Savoye M, Shaw M, Caprio S: Obesity dynamics and cardiovascular risk factor stability in obese adolescents. Pediatr Diabetes 2009;10: 360–367.

59 Savoye M, Shaw M, Dziura J, Tamborlane WV, Rose P, Guandalini C, Goldberg-Gell R, Burgert TS, Cali AM, Weiss R, Caprio S: Effects of a weight management program on body composition and metabolic parameters in overweight children: a randomized controlled trial. JAMA 2007;297:2697–2704.

60 Inge TH, Miyano G, Bean J, Helmrath M, Courcoulas A, Harmon CM, Chen MK, Wilson K, Daniels SR, Garcia VF, Brandt ML, Dolan LM: Reversal of type 2 diabetes mellitus and improvements in cardiovascular risk factors after surgical weight loss in adolescents. Pediatrics 2009;123: 214–222.

Ram Weiss, MD, PhD
Department of Human Nutrition and Metabolism, Braun School
of Public Health
Hebrew University School of Medicine, Jerusalem 91120 (Israel)
Tel. +972 2 643 1105, Fax +972 2 675 7070
E-Mail ram.weiss@ekmd.huji.ac.il

Levy-Marchal C, Pénicaud L (eds): Adipose Tissue Development: From Animal Models to Clinical Conditions.
Endocr Dev. Basel, Karger, 2010, vol 19, pp 73–83

Pathophysiology of Insulin Resistance in Small for Gestational Age Subjects: A Role for Adipose Tissue?

Jacques Beltrand · Taly Meas · Claire Levy-Marchal

INSERM, U690 and Université Paris 7 Denis Diderot, Paris, France

Abstract

Over the last 15 years a number of long-term health risks associated with reduced fetal growth have been identified, including cardiovascular diseases, hypertension, dyslipidemia, or type 2 diabetes. A common feature of these conditions is insulin resistance, which is thought to play a pathogenic role. However, despite abundant data in the literature, it is still difficult to trace the pathway by which fetal events, environmental or not, may lead to the increased morbidity later in life. To explain this association, several hypotheses have been proposed pointing to the role of either a detrimental fetal environment or a genetic susceptibility or an interaction between the two and of the particular dynamic changes in adiposity that occur during catch-up growth. The relative impact of early postnatal events in relation to fetal growth has to be considered for designing health policy strategies for early interventions aimed at decreasing the diseases risk throughout life.

Over the last 15 years, after the first observation of Barker et al. [1] in an adult population cohort, a number of long-term health risks for subjects born small for gestational age (SGA) have been identified, including hypertension, increased cardiovascular mortality [2], insulin resistance, impaired glucose tolerance, type 2 diabetes mellitus [3, 4], and dyslipidemia [4, 5]. A common feature of these conditions is the presence of high insulin levels, which are thought to play a pathogenic role. These important primary observations have led to the fetal origin hypothesis where fetal adaptation to an adverse intrauterine environment during a critical period of development may lead to long-lasting programming in a range of metabolic, physiological and structural parameters. However, inferences from data based on observational approaches require careful interpretation. Despite abundant data in the literature it is still difficult to trace the pathway by which fetal events, environmental or not, may lead to the increased morbidity later in life. Several cohorts of different ethnicities, ages, periods, lifestyle and dietary habits have then been studied in this respect.

Epidemiological Data

The very first observation made by Barker et al. [2] related early growth and death from ischemic heart diseases; from there the association was tested upstream with type 2 diabetes. Hales et al. [3] reported that a low birth weight was significantly associated with type 2 diabetes in a cohort of Caucasian males aged 64 years. Other epidemiological studies performed in Pima Indians or in Caucasian subjects later confirmed this observation, suggesting that the association holds true whether the study populations are genetically predisposed or not to type 2 diabetes [6, 7]. The independent effect of thinness at birth, assessed by the ponderal index, and the lack of association with gestational age on these complications observed in several studies strongly strengthen the deleterious effect of reduced fetal growth rather than prematurity on this association.

Low Birth Weight and Insulin Resistance

Evidence for Insulin Resistance in SGA Subjects
Insulin resistance associated with a small body size at birth was first mentioned on the basis of elevated insulin and proinsulin concentrations in 50-year-old subjects in whom birth weight was retrospectively collected [8]. The reality of insulin resistance was later assessed in children and in adults born SGA by the decreased peripheral glucose uptake measured by various methods such as IVGTT, minimal model and euglycemic hyperinsulinemic clamp [9–13]. Insulin sensitivity was reduced in SGA subjects irrespective of confounding factors, in particular the current body mass index (BMI) at the time of observation. In the Haguenau study, evidence of insulin resistance in SGA adults in comparison with appropriate for gestational age (AGA) subjects comes from indirect and direct measurements [11, 14, 15]; fasting insulin/glucose was significantly more elevated and QUICKI significantly decreased. When measured by the euglycemic hyperinsulinemic clamp in a subset of subjects, insulin sensitivity was decreased by around 20% in subjects born SGA. This insulin resistance was moderate in magnitude, independent from the usual factors associated with insulin resistance such as BMI, age, smoking, oral contraception and family history of diabetes and dyslipidemia and, importantly, did not affect all subjects born SGA but about a third of them.

Insulin Resistance in Subjects Born SGA Can Be Detected Early in Life

A number of studies have tested for the presence of insulin resistance in children born SGA. From these it is clear that insulin resistance appears very early in the postnatal period and most likely during the period of so-called 'catch-up' growth, between 0 and 2 years of life. What these studies also clearly demonstrate is that insulin resistance at this early age is very moderate indeed. For instance, studying 85 SGA babies as early as 1 year of age, Soto et al. [16] could not confirm a difference in insulin sensitivity in comparison to AGA babies. The only subjects in whom insulin resistance could be seen were SGAs with a catch-up growth or higher BMI. A similar observation was made in prepubertal SGA children in whom insulin sensitivity was measured with the clamp technique [13]. Insulin sensitivity was not affected in children born SGA with no catch-up and significantly affected in children with subsequent catch-up leading to a current BMI ≥ 17 kg/m^2. Glucose tolerance tests were also carried out on hundreds of 4-year-old children, whose birth weights were recorded; fasting and post-stimulation insulin concentrations were higher in individuals with a birth weight <2.4 kg [17]. It is interesting to emphasize that type 2 diabetes is frequent in India as well as low birth weight. Moreover, Indian neonates have been described as light but fat with central adiposity and hyperinsulinemia detectable at birth [18].

Altogether it seems that insulin resistance is a very early event in SGA babies prone to develop metabolic complications and would be amplified by an excess of fat mass. As an example, in a British study including around 600 10- to 11-year-old children, after adjustment for childhood height and ponderal index, fasting insulin levels fell with increasing birth weight. For each kilogram increase in birth weight, fasting insulin fell by 16.9%. However, the proportional change in insulin level for a 1 SD increase in childhood ponderal index was much greater than that for birth weight (27.2 and −8.8%, respectively, for fasting insulin) [19].

Postnatal Growth

The phenotype of lower birth weight associated with higher BMI in childhood and adulthood appears to predict the highest risks of insulin resistance and cardiovascular events. As described above, insulin resistance may be evidenced in SGA subjects very early in life, but clearly depends on their postnatal growth patterns. Accordingly, in a large longitudinal British cohort, lower birth weight was associated with reduced insulin sensitivity in childhood, but only in the highest current BMI tertile [20, 21]. By contrast, insulin secretion in these and other studies was directly related to the catch-up in height [13, 20]. Insulin is an important

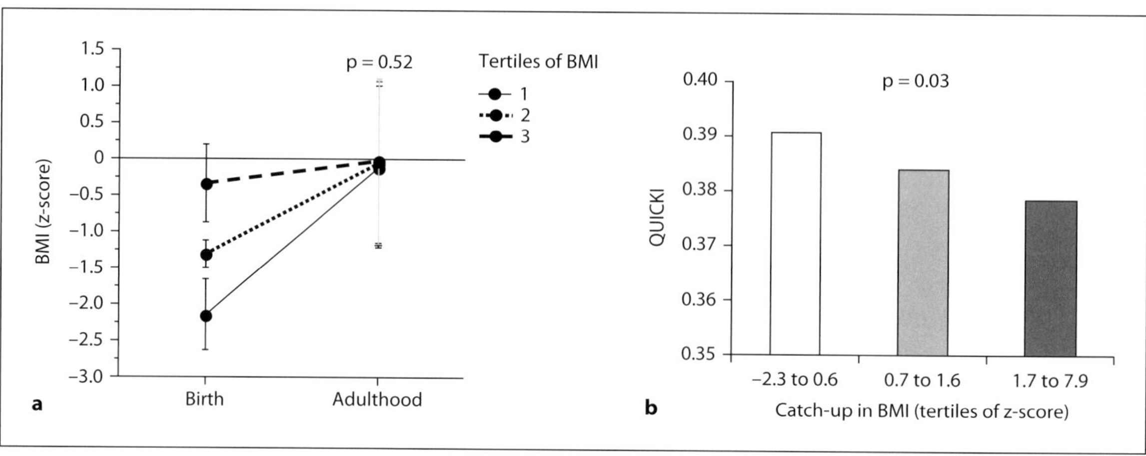

Fig. 1. Catch-up of adiposity and insulin resistance in subjects born SGA (n = 734). **a** Catch-up in BMI from birth to 22 years in subjects born SGA (n = 734). Subjects were divided into tertiles of BMI SDS at birth. Adult BMI is expressed in SDS according to gender. **b** Effect of catch-up in BMI SDS on the quantitative insulin sensitivity check index measured in adulthood. Catch-up in BMI SDS is expressed in tertiles.

growth factor during infancy, and higher secretion might lead to increased gain in length and higher IGF-I levels, as recently reported in 5-year-old children. In the Haguenau cohort, neither birth weight nor gestational age influenced insulin resistance in SGA subjects at the age of 22 years (fig. 1). By contrast, ponderal index at birth (kg/m³) was inversely related to the catch-up in BMI, which was in turn inversely associated with the insulin sensitivity index [11]. The association of postnatal weight gain with cardiovascular risk factors and diseases remains controversial as well. One issue that remains unresolved is the role of early postnatal growth from birth to 2 years of age. Studies in either preterm or full-term infants have described increased cardiovascular risk factors with rapid weight gain during early infancy [22]. In these studies, breast-feeding and consequently lower growth rates (in comparison with formula feeding which provides with higher nutrients and promotes early growth) were associated with lower levels of cholesterol in adolescence [23] and adulthood [24]. Similarly, the effect of early growth acceleration on insulin resistance was also substantial: for each z-score increase in neonatal weight gain, the concentration of 32–33 split proinsulin (a marker of insulin resistance) was 44% higher one to two decades later [25]. Furthermore, a recent randomized study compared the effect of a standard diet versus a nutrient-enriched diet promoting catch-up growth in SGA children in the first months of life. In observational analyses, catch-up growth induced by enriched diet was associated with a higher blood pressure at 7 years of age [26]. It should however be noted that conclusions drawn in these studies were based on intermediate outcomes, which are diseases risk but not clinically manifested diseases. In contrast, other observational studies on the historical cohorts have found coronary heart disease to be associated with lower weight at birth and during infancy and higher rates of weight gain after 2 years of age [27–29]. Some of the discrepancies between the studies may also result from differences in infants' feeding habits and growth in different time periods. Nonetheless, as a postnatal catch-up growth is characteristic for the vast majority of SGA infants, a lack of consensus regarding the impact of rapid early weight gain on adult health challenges clinicians and health professionals by questions on early feeding with energy-enriched formulas, as well as the need of effective strategies to promote the duration and exclusivity of breast-feeding. However, in preterm and SGA infants, potentially advantageous effects of relatively lower energy intake and slower growth in early life on long-term metabolic consequences and cardiovascular health must be considered bearing in mind the adverse effects of undernutrition on brain development [22] and retinopathy [30]. Consequently, further research is needed to clarify these points and for critical reappraisal of health policy strategies regarding recommendations for early feeding patterns.

Pathway to Insulin Resistance

Similar to the molecular mechanisms, the natural history of insulin resistance in SGA subjects is still being debated. It seems that insulin resistance would not occur at the same time in all insulin-dependent tissues. One study in young SGA and AGA (age 19 years) reported a normal whole-body insulin sensitivity (normal insulin-stimulated glucose disposal) but a decreased insulin-stimulated glycolytic flux [31]. This decreased insulin-stimulated glycolytic flux could represent an underlying primary defect at the cellular level, preceding and potentially leading to insulin resistance and diabetes later in life. In the same subjects, muscular expression of different molecules of insulin signalization pathway (PKC, p85α, p113β) and of GLUT4 was decreased in SGA subjects [32]. Another study reported reduced muscle glycolytic adenosine triphosphate production during exercise using ^{13}P-magnetic resonance spectroscopy in normoglycemic and insulin-resistant women who were thin at birth [33].

Taken together, these data emphasize that a defect in glycolysis could be the first step of the whole-body insulin resistance. It remains to be determined whether this defect is confined to skeletal muscle or if it is present in other tissues important for glucose homeostasis. Anyway, these data suggest that glucose could be redirected to other insulin-dependent tissue, such as adipose tissue, and would favor its growth leading to an excess of fat mass that would favor whole-body insulin resistance.

Role of Adipose Tissue

Pre- and Postnatal Adipose Tissue Development in SGA

Data from absorptiometry studies show that fat mass is indeed strikingly reduced in intrauterine growth-retarded neonates [34]. Thus, reduced fetal growth results in severely altered prenatal development of adipose tissue, which is mostly reflected by a decreased fat accumulation in adipocytes [35, 36]. It has also been shown that babies who are thin at birth lack muscle, a deficiency that persists into childhood and adulthood [37, 38]. It has therefore been suggested that thinness at birth may affect muscle structure and function and impair carbohydrate metabolism, as shown by reduced rates of glycolysis in muscle of adult subjects born with a low ponderal index. However, how it is linked to insulin resistance is unclear. Postnatal catch-up is a compensatory phenomenon to abnormal thinness or smallness at birth in order to bring each subject to his/her spontaneous 'genetic' centile. However, catch-up growth likely promotes an excess of adiposity in relation to muscle mass. In our cohort this is clearly illustrated by the similar and normal mean BMI in SGA and AGA young adults contrasting with a higher percent of fat mass in the SGA subjects (26.2 vs. 22.8% in AGA) [11]. In other situations of rapid and drastic wasting, the catch-up induces excessive fat deposition during nutritional rehabilitation, which in turn may promote selective insulin resistance [39]. Furthermore, in studies focusing on body composition and fat distribution, low birth weight has been associated with a more central pattern of fat distribution [40, 41]. In this perspective, insulin resistance associated with fetal growth restriction would be due to reshaping of body composition and impaired development of the adipose tissue independent of obesity.

Reduced Fetal Growth and Hormonal Function of the Adipose Tissue

Accumulating evidence points to the particular development of adiposity consecutive to reduced fetal growth associated with insulin resistance of the adipose tissue. Early insulin resistance of the adipose tissue can be evidenced while subjects show a preserved normal tolerance to glucose. Furthermore, in subjects born SGA, the decreased peripheral glucose uptake under the euglycemic hyperinsulinemic clamp closely correlates with an altered antilipolytic action of insulin assessed by the reduced suppression of free fatty acid production during the clamp [11]. Additionally, regulation of leptin production is altered in subjects born SGA both during the catch-up growth period and in adulthood [42]. Although the mechanism of this dysregulation remains to be clarified, these observations are of interest keeping in mind the role of leptin in the regulation of body fat mass and insulin sensitivity.

Similarly, a decreased adiponectin production and dysregulation in its insulin-sensitizing action have been observed in young adults born SGA in comparison with their peers born AGA [43]. In addition, adiponectin levels were found to be reduced in SGA as compared to AGA children, even lower than in obese children of the same age, and were inversely related to their weight gain [44, 45]. Furthermore, the abdominal subcutaneous tissue in SGA subjects shows a hyperlipolytic reactivity to catecholamines, which might be regarded as an additional deleterious component of insulin resistance associated with SGA [46]. Moreover, insulin resistance in adults is modulated by genetic polymorphisms of key players in the differentiation and insulin sensitivity of adipocytes such as PPARγ and β_3-adrenorecptors, but only in subjects of low birth weight [47, 48]. Taken together, these observations argue for alteration of the adipose tissue during the period of fetal growth restriction extending to the postnatal period with long-term functional consequences in adults.

Fat Growth and Insulin Resistance

Physiopathology of insulin resistance during catch-up growth is still being debated. Some hypotheses can be derived from mechanisms in other clinical situations of rapid weight gain. Increases in the ratio of fat mass to lean mass are well documented in adults recovering body weight after a weight loss due to a variety of conditions such as experimental starvation, weight loss in obese, or war-related famine. Thus a common denominator in many situations where there are large decreases in body weight followed by weight recovery (whether during growth or in adulthood) is that body fat is recovered at a disproportionately faster rate than lean tissue. Explanations for this phenomenon of preferential catch-up fat have, in the past, centered upon inadequate intake of dietary protein or other nutrients for optimal protein deposition [48]. However, the fact that catch-up fat persists in patients on well-balanced diets and independently of the level of dietary energy underscores an increase in metabolic efficiency directed as fat deposition as a fundamental physiological process operating to accelerate fat recovery after growth retardation or weight loss. This phenomenon of catch-up fat could be a normal physiological process characterized by an elevation in the efficiency of cellular utilization.

It has been shown during experimental starvation that fat depletion was associated with a reduced basal metabolic rate (fig. 2). The role would be to specifically accelerate body fat recovery as reported by animal studies. If there is a reduced basal metabolic rate, stored energy would be specifically directed for the recovery of fat mass rather than of lean tissue. This 'adipose-specific control of energy expenditure' could operate as a loop between adipose tissue triglyceride stores

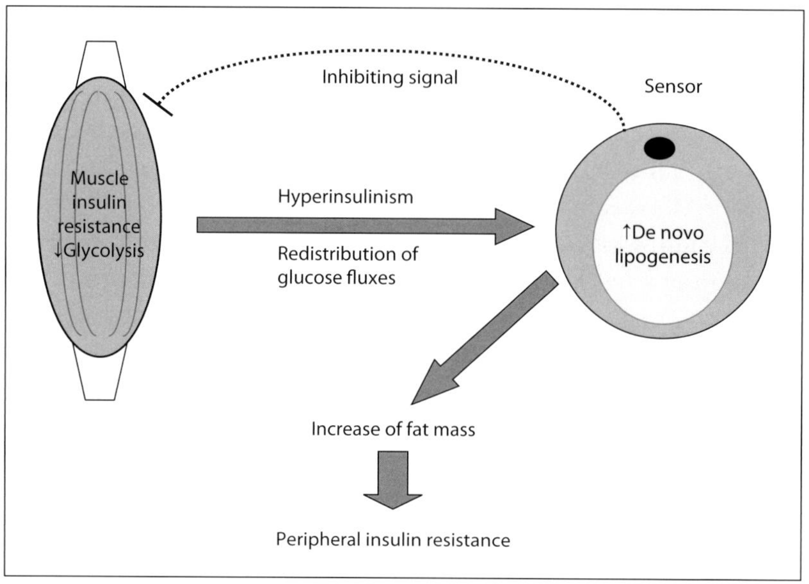

Fig. 2. Development of insulin resistance during catch-up growth, a schematic view.

and skeletal muscle metabolism. Specific muscular insulin resistance and reduction in the metabolic rate of muscle would result in a reduction of glucose utilization leading to hyperinsulinemia. This in turn would redirect the spared glucose towards de novo lipogenesis and fat storage in adipose tissue. Such a phenomenon has been reported in animals where the state of muscle insulin resistance is associated to a state of hyperinsulin sensitivity in the adipocytes enhancing conversion of glucose to lipids in fat stores [49–51].

Despite its adaptive nature within the context of a lifestyle of famine and feast, this state may have deleterious consequences in the context of the modern lifestyle. In our context of complex interaction between earlier reprogramming and a modern lifestyle, this thrifty catch-up fat phenotype is a central event that predisposes individuals with catch-up growth to type 2 diabetes and cardiovascular diseases.

Conclusion

Accumulating evidence from numerous studies worldwide is now conclusive that small size at birth confers an increased risk for metabolic and cardiovascular disorders in later life. Insulin resistance is a keystone for the simultaneous development of the metabolic syndrome and the later occurrence of type 2 diabetes and it is amplified by obesity in adulthood. Clearly, not all subjects born SGA are at the same risk of developing these complications. It appears that individuals especially at risk are those who were thin at birth and had subsequent catch-up in BMI, irrespective of the degree of adiposity. It is suggested that this particular dynamic change in adiposity has a critical role in the development of long-term metabolic complications. Since thinness at birth encompasses many identified or unidentified causes, it remains to be clarified whether the mechanisms responsible for the long-term metabolic outcome are specific to some etiologies or result from the dynamic process leading to thinness itself. Another key issue is the relative importance of the early postnatal events to the diseases risk throughout life in forming health policy strategies towards eventual early interventions in addition to those exerted in adulthood. Lack of consensus on the impact of early postnatal nutrition and weight gain on later metabolic outcomes underlines the need of further research in this field before conclusive recommendations of early nutrition will be appropriate.

References

1   Barker DJ, Osmond C, Golding J, Kuh D, Wadsworth ME: Growth in utero, blood pressure in childhood and adult life, and mortality from cardiovascular disease. BMJ 1989;298:564–567.

2   Barker DJ, Winter PD, Osmond C, Margetts B, Simmonds SJ: Weight in infancy and death from ischaemic heart disease. Lancet 1989;2:577–580.

3   Hales CN, Barker DJ, Clark PM, Cox LJ, Fall C, Osmond C, Winter PD: Fetal and infant growth and impaired glucose tolerance at age 64. BMJ 1991;303:1019–1022.

4   Barker DJ, Hales CN, Fall CH, Osmond C, Phipps K, Clark PM: Type 2 (non-insulin-dependent) diabetes mellitus, hypertension and hyperlipidaemia (syndrome X): Relation to reduced fetal growth. Diabetologia 1993;36:62–67.

5   Fall CH, Barker DJ, Osmond C, Winter PD, Clark PM, Hales CN: Relation of infant feeding to adult serum cholesterol concentration and death from ischaemic heart disease. BMJ 1992;304:801–805.

6   Lithell HO, McKeigue PM, Berglund L, Mohsen R, Lithell UB, Leon DA: Relation of size at birth to non-insulin-dependent diabetes and insulin concentrations in men aged 50–60 years. BMJ 1996;312:406–410.

7   McCance DR, Pettitt DJ, Hanson RL, Jacobsson LT, Knowler WC, Bennett PH: Birth weight and non-insulin-dependent diabetes: thrifty genotype, thrifty phenotype, or surviving small baby genotype? BMJ 1994;308:942–945.

8   Phipps K, Barker DJ, Hales CN, Fall CH, Osmond C, Clark PM: Fetal growth and impaired glucose tolerance in men and women. Diabetologia 1993; 36:225–228.

9   Flanagan DE, Moore VM, Godsland IF, Cockington RA, Robinson JS, Phillips DI: Fetal growth and the physiological control of glucose tolerance in adults: a minimal model analysis. Am J Physiol 2000;278:E700–E706.

10 Hofman PL, Cutfield WS, Robinson EM, Bergman RN, Menon RK, Sperling MA, Gluckman PD: Insulin resistance in short children with intrauterine growth retardation. J Clin Endocrinol Metab 1997;82:402–406.

11 Jaquet D, Gaboriau A, Czernichow P, Levy-Marchal C: Insulin resistance early in adulthood in subjects born with intrauterine growth retardation. J Clin Endocrinol Metab 2000;85:1401–1406.

12 Phillips DI, Barker DJ, Hales CN, Hirst S, Osmond C: Thinness at birth and insulin resistance in adult life. Diabetologia 1994;37:150–154.

13 Veening MA, Van Weissenbruch MM, Delemarre-Van De Waal HA: Glucose tolerance, insulin sensitivity, and insulin secretion in children born small for gestational age. J Clin Endocrinol Metab 2002;87:4657–4661.

14 Jaquet D, Deghmoun S, Chevenne D, Collin D, Czernichow P, Levy-Marchal C: Dynamic change in adiposity from fetal to postnatal life is involved in the metabolic syndrome associated with reduced fetal growth. Diabetologia 2005;48:849–855.

15   Leger J, Levy-Marchal C, Bloch J, Pinet A, Chevenne D, Porquet D, Collin D, Czernichow P: Reduced final height and indications for insulin resistance in 20-year-olds born small for gestational age: regional cohort study. BMJ 1997;315: 341–347.

16   Soto N, Bazaes RA, Pena V, Salazar T, Avila A, Iniguez G, Ong KK, Dunger DB, Mericq MV: Insulin sensitivity and secretion are related to catch-up growth in small-for-gestational-age infants at age one year: results from a prospective cohort. J Clin Endocrinol Metab 2003;88:3645–3650.

17   Yajnik CS, Fall CH, Vaidya U, Pandit AN, Bavdekar A, Bhat DS, Osmond C, Hales CN, Barker DJ: Fetal growth and glucose and insulin metabolism in four-year-old Indian children. Diabet Med 1995;12:330–336.

18   Yajnik CS, Lubree HG, Rege SS, Naik SS, Deshpande JA, Deshpande SS, Joglekar CV, Yudkin JS: Adiposity and hyperinsulinemia in Indians are present at birth. J Clin Endocrinol Metab 2002;87: 5575–5580.

19   Whincup PH, Cook DG, Adshead F, Taylor SJ, Walker M, Papacosta O, Alberti KG: Childhood size is more strongly related than size at birth to glucose and insulin levels in 10- to 11-year-old children. Diabetologia 1997;40:319–326.

20   Ong KK, Petry CJ, Emmett PM, Sandhu MS, Kiess W, Hales CN, Ness AR, Dunger DB: Insulin sensitivity and secretion in normal children related to size at birth, postnatal growth, and plasma insulin-like growth factor-I levels. Diabetologia 2004;47:1064–1070.

21   Colle E, Schiff D, Andrew G, Bauer CB, Fitzhardinge P: Insulin responses during catch-up growth of infants who were small for gestational age. Pediatrics 1976;57:363–371.

22   Singhal A, Lucas A: Early origins of cardiovascular disease: Is there a unifying hypothesis? Lancet 2004;363:1642–1645.

23   Singhal A, Cole TJ, Fewtrell M, Lucas A: Breast milk feeding and lipoprotein profile in adolescents born preterm: follow-up of a prospective randomised study. Lancet 2004;363:1571–1578.

24   Owen CG, Whincup PH, Odoki K, Gilg JA, Cook DG: Infant feeding and blood cholesterol: a study in adolescents and a systematic review. Pediatrics 2002;110:597–608.

25   Singhal A, Fewtrell M, Cole TJ, Lucas A: Low nutrient intake and early growth for later insulin resistance in adolescents born preterm. Lancet 2003;361:1089–1097.

26   Lawlor DA, Hubinette A, Tynelius P, Leon DA, Smith GD, Rasmussen F: Associations of gestational age and intrauterine growth with systolic blood pressure in a family-based study of 386,485 men in 331,089 families. Circulation 2007;115: 562–568.

27   Barker DJ, Osmond C, Forsen TJ, Kajantie E, Eriksson JG: Trajectories of growth among children who have coronary events as adults. N Engl J Med 2005;353:1802–1809.

28   Eriksson JG, Forsen T, Tuomilehto J, Osmond C, Barker DJ: Early growth and coronary heart disease in later life: longitudinal study. BMJ 2001; 322:949–953.

29   Forsen TJ, Eriksson JG, Osmond C, Barker DJ: The infant growth of boys who later develop coronary heart disease. Ann Med 2004;36:389–392.

30   Engstrom E, Niklasson A, Wikland KA, Ewald U, Hellstrom A: The role of maternal factors, postnatal nutrition, weight gain, and gender in regulation of serum IGF-I among preterm infants. Pediatr Res 2005;57:605–610.

31   Jensen CB, Storgaard H, Dela F, Holst JJ, Madsbad S, Vaag AA: Early differential defects of insulin secretion and action in 19-year-old Caucasian men who had low birth weight. Diabetes 2002;51:1271–1280.

32   Ozanne SE, Jensen CB, Tingey KJ, Storgaard H, Madsbad S, Vaag AA: Low birth weight is associated with specific changes in muscle insulin-signalling protein expression. Diabetologia 2005;48: 547–552.

33   Taylor DJ, Thompson CH, Kemp GJ, Barnes PR, Sanderson AL, Radda GK, Phillips DI: A relationship between impaired fetal growth and reduced muscle glycolysis revealed by ^{31}P magnetic resonance spectroscopy. Diabetologia 1995;38:1205–1212.

34   Lapillonne A, Braillon P, Claris O, Chatelain PG, Delmas PD, Salle BL: Body composition in appropriate and in small for gestational age infants. Acta Paediatr 1997;86:196–200.

35   Enzi G, Zanardo V, Caretta F, Inelmen EM, Rubaltelli F: Intrauterine growth and adipose tissue development. Am J Clin Nutr 1981;34:1785–1790.

36   Petersen S, Gotfredsen A, Knudsen FU: Lean body mass in small for gestational age and appropriate for gestational age infants. J Pediatr 1988; 113:886–889.

37   Eriksson J, Forsen T, Tuomilehto J, Osmond C, Barker D: Size at birth, fat-free mass and resting metabolic rate in adult life. Horm Metab Res 2002;34:72–76.

Beltrand · Meas · Levy-Marchal

38 Hediger ML, Overpeck MD, Kuczmarski RJ, McGlynn A, Maurer KR, Davis WW: Muscularity and fatness of infants and young children born small- or large-for-gestational-age. Pediatrics 1998;102:E60.

39 Dulloo AG: Regulation of fat storage via suppressed thermogenesis: a thrifty phenotype that predisposes individuals with catch-up growth to insulin resistance and obesity. Horm Res 2006;65 (suppl 3):90–97.

40 Barker M, Robinson S, Osmond C, Barker DJ: Birth weight and body fat distribution in adolescent girls. Arch Dis Child 1997;77:381–383.

41 Rasmussen EL, Malis C, Jensen CB, Jensen JE, Storgaard H, Poulsen P, Pilgaard K, Schou JH, Madsbad S, Astrup A, Vaag A: Altered fat tissue distribution in young adult men who had low birth weight. Diabetes Care 2005;28:151–153.

42 Jaquet D, Gaboriau A, Czernichow P, Levy-Marchal C: Relatively low serum leptin levels in adults born with intrauterine growth retardation. Int J Obes Relat Metab Disord 2001;25:491–495.

43 Jaquet D, Deghmoun S, Chevenne D, Czernichow P, Levy-Marchal C: Low serum adiponectin levels in subjects born small for gestational age: Impact on insulin sensitivity. Int J Obes (Lond) 2006;30: 83–87.

44 Cianfarani S, Martinez C, Maiorana A, Scire G, Spadoni GL, Boemi S: Adiponectin levels are reduced in children born small for gestational age and are inversely related to postnatal catch-up growth. J Clin Endocrinol Metab 2004;89:1346–1351.

45 Iniguez G, Soto N, Avila A, Salazar T, Ong K, Dunger D, Mericq V: Adiponectin levels in the first two years of life in a prospective cohort: Relations with weight gain, leptin levels and insulin sensitivity. J Clin Endocrinol Metab 2004;89:5500–5503.

46 Boiko J, Jaquet D, Chevenne D, Rigal O, Czernichow P, Levy-Marchal C: In situ lipolytic regulation in subjects born small for gestational age. Int J Obes (Lond) 2005;29:565–570.

47 Eriksson JG, Lindi V, Uusitupa M, Forsen TJ, Laakso M, Osmond C, Barker DJ: The effects of the Pro12Ala polymorphism of the peroxisome proliferator-activated receptor-γ_2 gene on insulin sensitivity and insulin metabolism interact with size at birth. Diabetes 2002;51:2321–2324.

48 Jaquet D, Tregouet DA, Godefroy T, Nicaud V, Chevenne D, Tiret L, Czernichow P, Levy-Marchal C: Combined effects of genetic and environmental factors on insulin resistance associated with reduced fetal growth. Diabetes 2002;51:3473–3478.

49 Dulloo AG, Jacquet J: An adipose-specific control of thermogenesis in body weight regulation. Int J Obes Relat Metab Disord 2001;25(suppl 5):S22–S29.

50 Dulloo AG: A role for suppressed skeletal muscle thermogenesis in pathways from weight fluctuations to the insulin resistance syndrome. Acta Physiol Scand 2005;184:295–307.

51 Cettour-Rose P, Samec S, Russell AP, Summermatter S, Mainieri D, Carrillo-Theander C, Montani JP, Seydoux J, Rohner-Jeanrenaud F, Dulloo AG: Redistribution of glucose from skeletal muscle to adipose tissue during catch-up fat: a link between catch-up growth and later metabolic syndrome. Diabetes 2005;54:751–756.

Claire Levy-Marchal
INSERM, U690, Hôpital Robert Debré
48 boulevard Sérurier, FR–75019 Paris (France)
Tel. +33 140 03 19 86, Fax +33 140 40 91 96
Mail claire.levy-marchal@inserm.fr

Levy-Marchal C, Pénicaud L (eds): Adipose Tissue Development: From Animal Models to Clinical Conditions.
Endocr Dev. Basel, Karger, 2010, vol 19, pp 84–92

The Neural Feedback Loop between the Brain and Adipose Tissues

Luc Pénicaud

UMR 6265, CNRS Université de Bourgogne, Dijon, France

Abstract

There are more and more data supporting the importance of nervous regulation of both white and brown adipose tissue mass. This short paper will review the different physiological parameters which are regulated such as metabolism (lipolysis and thermogeneis), secretory activity (leptin and other adipokines) but also to plasticity of adipose tissues (proliferation differentiation and apoptosis). The sensory innervation of white adipose issue and its putative role will be also described. Altogether these results showed the presence of a neural feedback loop between adipose tissues and the brain which plays a major role in the regulation of energy homeostasis and has been shown to be altered in physiologic as well as in metabolic pathologies.   Copyright © 2010 S. Karger AG, Basel

Factors of neural origin play an important role in the control of energy homeostasis. Indeed central and autonomic nervous systems are involved in the regulation of whole-body energy by regulating its different components: intake, expenditure and storage. Among the tissues of which the activity is controlled by the autonomic nervous system, adipose tissues play a crucial role both via their metabolic and secretory capacities.

In most mammals, two types of adipose tissue, white and brown, are present. Both tissues are able to store energy in the form of triacylglycerols and to hydrolyze them into free fatty acids and glycerol. Whereas white adipose tissue (WAT) provide lipids as substrates for other tissues according to the needs of the organism, brown adipose tissue (BAT) uses fatty acids for heat production. Over a period of time, white fat mass reflects the balance between energy expenditure and energy intake. Remarkably, body fat mass remains relatively constant in adults suggesting that food intake and energy expenditure are linked. This has been supported by numerous studies that demonstrated the interdependency of these parameters and thus a feedback loop between the brain and adipose tissues with the involvement of the autonomic nervous system and that of sensory fibers.

From the Brain to Adipose Tissues

Sympathetic Nervous System (SNS)
It has long been recognized that BAT receives dense SNS innervation both at the level of blood vessels but also directly on adipocytes [1]. It is believed that each brown adipocyte receives one or more nerve endings [2]. In WAT, the noradrenergic innervation fibers have initially been reported as closely associated with the blood vessels leading to its implication in WAT blood flow [3, 4]. Whereas data questioned this in the late 1980s [5], it was in 1995 that Youngstrom and Bartness [6] demonstrated direct neuroanatomical evidence for innervation of white adipocytes. Furthermore, they elegantly showed, using viral tracing methodologies, that the inputs originate from CNS nuclei (paraventricular hypothalamus, noradrenergic tegmental system, caudal raphe region, etc.) that are part of the SNS outflow [7, 8].

The neurotransmitter involved is mainly norepinephrine that binds to different noradrenergic receptor subtypes [9]. In BAT, β-adrenergic receptor subtype (β_3 in rodents) is the only one present and lipolysis is always activated when the sympathetic drive increases [10]. In WAT it has been demonstrated by Lafontan and Berlan [11] that the lipolytic activity of adipocytes depends on a balance between lipolysis-promoting β-adrenergic receptor and lipolysis-inhibiting α_2-adrenergic receptor. Thus depending on this balance, an increased sympathetic tone can lead to an increase or a decrease in lipolysis.

One has also to mention that in addition to norepinephrine, nerves also contain neuropeptide such as neuropeptide Y [12, 13].

Parasympathetic Nervous System (PNS)
As many tissues are innervated by both SNS and PNS, evidence of PNS innervation of adipose tissues has been looked for. Recent neuroanatomical studies in rats have reported parasympathetic innervation of WAT. A physiological role of such an input was proposed since vagotomy was shown to reduce the insulin-dependent glucose and free fatty acid uptakes [14]. Such a role of PNS can be also sustained by the demonstration of the presence of functional nicotinic receptor on white adipocytes as well as an increased insulin sensitivity of these cells under nicotine stimulation [15]. However, the PNS innervation of WAT is highly controversial and remains a subject of debates [16–18].

Heterogeneity of WAT Autonomic Innervation
Although adipose tissue is used as a general term, the fat pads are quite different with regard to their origin, anatomical characteristics and functions so that one should rather speak about adipose tissues. Indeed both autonomic innervation and the number and affinity of the adrenoceptors of fat depots are heterogeneous. First a relatively separated sympathetic innervation of inguinal and epididymal

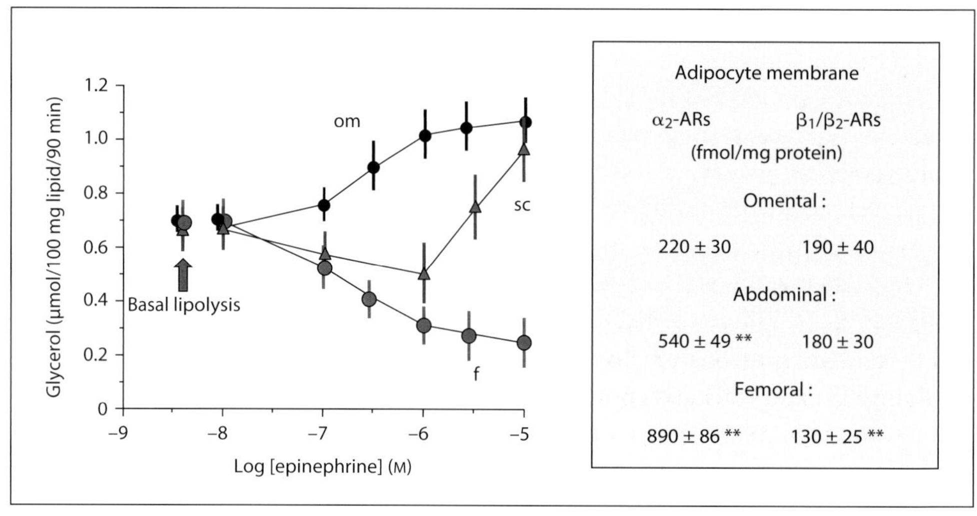

Fig. 1. Lipolytic response and number of adrenergic receptors (α_2 and β_1-β_2 subtypes) of adipocytes isolated from different fat localizations (omental (om), abdominal (sc) and femoral (f)) in non-obese women [adapted from 21].

pads exist, since there are no overlapping patterns of labeled postganglionic cells within the sympathetic chain innervating these two deposits using fluorescent tracers [6]. Viral tracer technologies also suggest differential innervation of fat pads for both sympathetic and parasympathetic nerves [7, 8, 14, 19]. Second, taking norepinephrine turnover as an index of SNS activity, a specific pattern has been delineated which might also depend of the stimulus considered [20]. Altogether these latter data indicate a higher lipolysis in intra-abdominal fat pads as compared to subcutaneous ones. Third, this is reinforced by the distribution of the different subclasses of receptors that depends on species, sex, and fat depot. Thus, in women, for example, it has been demonstrated that the number of α_2 and β_1, β_2-adrenoceptors varies between omental, abdominal and femoral adipose tissues and as a consequence the lipolytic response to epinephrine or norepinephrine [see above and 11, 21–23] (fig 1).

Role of Autonomic Nervous System in Adipose Tissues Biology

Metabolism
The main metabolic pathways of both white and brown adipocytes are the synthesis and accumulation of triacylglycerols and their degradation into free fatty acid and glycerol [24]. It is now recognized that lipolytic pathways are mainly under

the dependency of three main players: adipose triglyceride lipase, hormone-sensitive lipase, and perilipin A [25]. With regard to metabolism, the main differences between white and brown adipocytes is located downstream to lipolysis. Indeed in white adipocyte, both free fatty acids and glycerol are released into the adjacent blood vessels to provide fuel for other tissues. In brown adipocyte, free fatty acids will be oxidized by the cell itself. The energy produced will be then dissipated as heat, thanks to the presence of a specific mitochondrial protein: the uncoupling protein-1 [26]. As already mentioned above, catecholamines and particularly norepinephrine are the main hormones involved in the control of lipolysis in both cell types. However, one has to underline that the anti-lipolytic effect of insulin is predominant and thus catecholamines exert their effect when the insulin level is low. From what is said above it is easy to conclude that the SNS is the main driver for adipose tissues lipolysis.

In addition to this effect, the SNS plays a role in regulating anabolic pathways. It has been demonstrated that sympathetic nerves control glucose uptake, utilization and lipogenesis in BAT but not in WAT [27–29], whereas, as already mentioned, there is evidence that PNS innervation increases insulin sensitivity in WAT [14].

Secretion

Over the last 20 years the notion has emerged that WAT is not only involved in the storage and release of energy but could also be part of other physiological functions due to its capabilities in synthesis and secretion of numerous factors such as leptin, adiponectin and many proteins involved in inflammation and immunity [28, 29], so that adipose tissue is now considered as a true endocrine organ.

The synthesis and secretion of some of these compounds are under the control of numerous factors among which the SNS via catecholamines plays a role. Leptin control has probably been the most studied. There is ample evidence that stimulation of β-adrenoceptor decreases the release of leptin. In human adipose tissue this occurs through a posttranslational mechanism, most likely secretion per se. In contrast, in rat adipose tissue, isoproterenol does not affect basal leptin secretion but has a short-term action to antagonize the insulin-stimulated leptin biosynthesis [30, 31], although an elegant study demonstrates a decrease in leptin secretion when 3T3L1 adipocytes (a well-characterized white adipose cell line) are cultured in the presence of primary sympathetic neurons [32]. It has also been proposed that catecholamines may mediate a short-term decrease in plasma leptin that occurs within hours of fasting and cold exposure [33].

Adiponectin is also negatively regulated by β-adrenoceptor [34]. In contrast the secretion of cytokines such as TNF-α and IL-6 are increased under β-adrenergic stimulation [34–36]. Overall these data suggest that upregulation of proinflammatory cytokines and downregulation of adiponectin by β-adrenoceptor activation may contribute to the pathogenesis of catecholamine-induced insulin resistance.

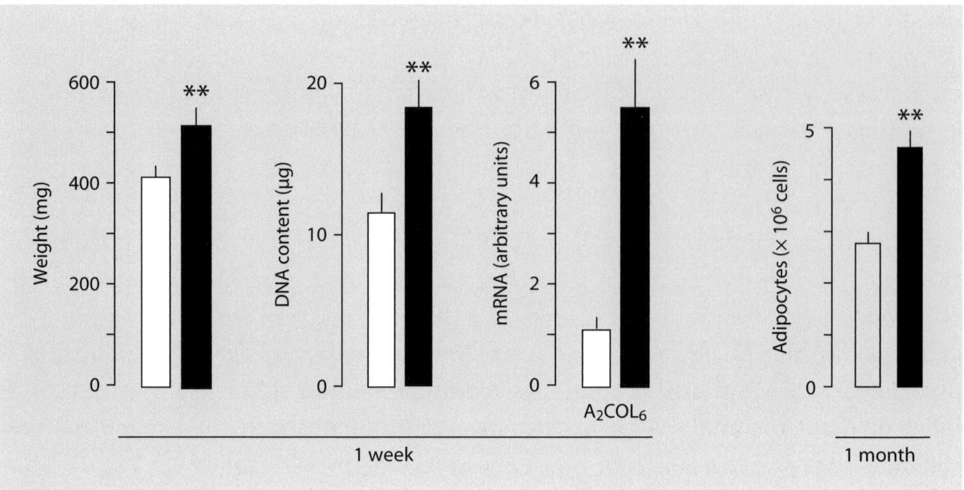

Fig. 2. Fat pad weight, DNA content, expression of A_2COL_6 and number of mature adipocytes observed in denervated (black column) or control (white column) retroperitoneal fat pads, 1 week or 1 month after surgery, in rats. ** p < 0.01.

Proliferation Differentiation and Apopotosis

Fat mass is the result of two processes, i.e. the regulation of the size and the number of adipocytes. We have shown that the autonomic nervous system is indeed involved in the first one by regulating both energy stores and thermogenesis in WAT and BAT respectively. There is also ample evidence showing that the SNS is involved in the control of proliferation and differentiation, and to a lesser of apoptosis of white and brown adipocytes.

It is then well established than norepinephrine induces both proliferation and differentiation of brown adipocytes precursors in vivo and in vitro [37–39]. Results observed after a decreased (denervation, noradrenergic blockade, hypothalamic lesion) or increased (β-agonist treatment, cold exposure) sympathetic activity lead to the same general conclusion.

In contrast, sympathetic activation would inhibit the development of WAT [23]. Norepinephrine inhibits proliferation of adipocyte precursor cells in vitro and can be blocked by propranolol, a general β-adrenoceptor antagonist [40]. In vivo surgical denervation of WAT triggers significant increases both in rats and Siberian hamsters [8, 41, 42]. Thus we have been the first to demonstrate that 1 week after denervation of one retroperitoneal fat pad, DNA content was largely increased without change in the number of mature white adipocytes. Furthermore, the amount of A_2COL_6, an early marker of white adipocyte differentiation, was enhanced in the denervated pad. One month later, the number of mature adipocytes was significantly increased in the denervated pad (fig. 2) [41]. A recent study using transgenic mice having a massive reduction of innervation due to the lack of Nscl-2, a neuronal-specific transcription factor, came in support of such an observation [43]. These mice present an increased preadipocytes number and a bimodal distribution of the size of adipocytes indicating an increase in the number of small adipocytes.

Although the importance of apoptosis in the biology of adipose tissues is still a controversial issue, there are different reports describing such a process in both white and brown adipocytes [23]. To our knowledge there is no direct demonstration of a role of the SNS in regulating the rate of apoptosis in adipose tissues, however several observations are in support of such a role. Leptin and insulin injected either at the periphery or centrally induce a reduction of fat pad weight that is associated with apoptosis [44–46]. Since both hormones enhance SNS activity, it is believed that the signal that promotes apoptosis under insulin and leptin CNS activation is probably norepinephrine or another co-secreted neurotransmitter [23].

Together, these results demonstrate that in vivo SNS innervation of WAT and BAT acts as modulator of fat cell development.

From the Adipose Tissues to the Brain

Sensory Innervation

To explain the precise regulation of energy balance and thus body fat mass, hypotheses have been proposed suggesting that signals generated in proportion to body fat stores will act on the brain to modulate food intake and/or energy expenditure. Among these signals, insulin and more importantly leptin have been demonstrated to play a crucial role [23, 46, 47]. Apart from hormones and metabolites, adipose tissues sensory nerves may be part of this system. Indeed sensory innervation of WAT has been demonstrated by various facts. The identification of substance P and calcitonin gene-related peptide, markers of sensory neurons, was a first demonstration [13]. Then a direct neuroanatomical demonstration was given by use of anterograde tracer [48]. Finally the sensory projection to different brain areas was extensively studied by Bartness and colleagues [20, 49]. As stated by these authors, 'labelling cells were found at all levels of the neuroaxis including both the nodose ganglia (visceral afferents) as well as the dorsal horn (spinal afferents) of the spinal cord and in almost all the autonomic output areas in the brainstem and midbrain'.

Although one does not know what (leptin, lipid molecules such as glycerol, free fatty acids, prostaglandins) these nerves 'sense', data are in support of their role in informing the brain on lipid stores. When selective destruction of sensory innervating epididymal fat pad was performed in hamsters by injecting capsaicin in one pad, the weight of the contralateral non-injected pad was increased in a degree that approximated the lipid deficit if the pad had been removed by lipectomy [50, 51].

Perspectives

There are more and more data supporting the importance of nervous regulation of adipose tissue mass, either brown or white, with regard not only to metabolic pathways but also to the plasticity of these tissues by regulating proliferation differentiation and apoptosis. The sensory innervation is a new field that merits further attention. Altogether this neural feedback loop between adipose tissues and the brain plays a role in the regulation of energy homeostasis and has been shown to be altered in obesity and type 2 diabetes [23]. Considering plasticity, more data are needed to strongly sustain the influence of the SNS on apoptosis. We also need further research on the importance of parasympathetic innervation. Finally, one issue of great interest in the general field of biology and development would be to determine whether nervous control could play a role in the fate of the different progenitors (stem cells, multipotent cells), namely adipose-derived stem cells which have been demonstrated to be present in adipose tissues [52–54] and of which the role, the differentiation potential and their regulation is still poorly understood.

References

1   Norman D, Mukherjee S, Symons D, Jung RT, Lever JD: Neuropeptides in interscapular and perirenal brown adipose tissue in the rat: a plurality of innervation. J Neurocytol 1988;17:305.

2   Himms-Hagen J: Brown adipose thermogenesis and obesity. Front Neuroendocrinol 1991;12:38.

3   Ballantyne B, Raffery AT: The intrinsic autonomic innervation of white adipose tissue. Cytobios 1974;10:187.

4   Slavin BG, Ballard KW: Morphological studies of the adrenergic innervation of white adipose tissue. Anat Rec 1978;191:377.

5   Rebuffé-Scrive M: Neuroregulation of adipose tissue: molecular and hormonal mechanisms. Int J Obes 1991;15:83.

6   Youngstrom TG, Bartness TJ: Catecholaminergic innervation of white adipose tissue in the Siberian hamster. Am J Physiol 1995;268:R744.

7   Bamshad M, Aoki VT, Adkison MG, Warren WS, Bartness TJ: Central nervous system origins of the sympathetic system outflow to white adipose tissue. Am J Physiol 1998;276:R291.

8   Bowers RR, Festuccia WT, Song CK, Shi, H, Migliorini RH, Bartness TJ: Sympathetic innervation of adipose tissue and its regulation of fat cell number. Am J Physiol 2004;286:R1167.

9   Lafontan M, Berlan M: Fat cell adrenergic receptor and the control of white and brown fat cell function. J Lipid Res 1993;34:1057.

10 Grujic D, Susulic VS, Harper ME, Himms-Hagen J, Cunningham BA, Corkey BE, Lowell BB: β_3-Adrenergic receptors on white and brown adipocytes mediate β_3-selective agonist-induced effects on energy expenditure, insulin secretion, and food intake. A study using transgenic and gene knockout mice. J Biol Chem 1997;272: 17686.

11 Lafontan M, Berlan M: Fat cell α_2-adrenoceptors: the regulation of fat cell function and lipolysis. Endocr Rev 1995;16:716.

12 Potter K: Neuropeptide Y as an autonomic neurotransmitter. Pharmacol Ther 1988;37:251.

13 Giordano A, Morroni M, Santone G, Marchesi GF, Cinti S: Tyrosine hydroxylase, neuropeptide Y, substance P, calcitonin gene-related peptide and vasoactive intestinal peptide in nerves of rat periovarian adipose tissue: an immunohistochemical and ultrastructural investigation. J Neurocytol 1996;25:125–136.

14 Kreier F, Fliers E, Voshol PJ, Van Eden CG, Havekes LM, Kalsbeek A, Van Heijningen CL, Sluiter AA, Mettenleiter TC, Roijn JA, Sauerwein HP, Buijs RM: Selective parasympathetic innervation of subcutaneous and intra-abdominal fat-functional implications. J Clin Invest 2002;110: 1243.

15 Liu RH, Mizuta M, Matsukura S: The expression and functional role of nicotinic acetylcholine receptors in rat adipocytes. J Pharmacol Exp Ther 2004;310:52.

16 Berthoud HR, Fox EA, Neuhuber WL: Controversial white adipose tissue innervation. Am J Physiol 2007;293:R553.

17 Giordano A, Song CK, Bowers RR, Ehlen JC, Frontini A, Cinti S, Bartness TJ: Hite adipose tissue lacks significant vagal innervation and immunohistochemical evidence of parasympathetic innervation. Am J Physiol 2006;291:R1243.

18 Kreier F, Buijs RM: Evidence for parasympathetic innervation of white adipose tissue, clearing up some vagaries. Am J Physiol 2007;293:R548.

19 Kreier F, Kap YS, Mettenleiter TC, van Heijningen C, van der Vliet J, Kalsbeek A, Sauerwein HP, Fliers E, Romijn JA, Buijs RM: Tracing from fat tissue, liver, and pancreas: a neuroanatomical framework for the role of the brain in type 2 diabetes. Endocrinology 2006;147:1140.

20 Bartnes TJ, Shrestha Y, Vaughan CH, Schwartz GJ, Song CK: Sensory and sympathetic nervous system control of white adipose tissue lipolysis. Mol Cell Endocrinol 2009.

21 Mauriege P, Galitzky J, Berlan M, Lafontan M: Heterogeneous distribution of β- and α_2- adrenoceptor binding sites in human fat cells from various fat deposits: functional consequences. Eur J Clin Invest 1987;17:156.

22 Mauriège P, De Pergola G, Berlan M, Lafontan M: Human fat cell β-adrenergic receptors: β-agonist-dependent lipolytic responses and characterization of β-adrenergic binding sites on human fat cell membranes with highly selective β_1-antagonists. J Lipid Res 1988;29:587.

23 Pénicaud L, Cousin B, Leloup C, Lorsignol A, Casteilla L: The autonomic nervous system, adipose tissue plasticity and energy balance. Nutrition 2000;16:903.

24 Wang S, Soni KG, Semache M, Casavant S, Fortier M, Pan L, Mitchell GA: Lipolysis and the integrated physiology of lipid energy metabolism. Mol Genet Metab 2008;95:117.

25 Ricquier D, Bouillaud F: Mitochondrial uncoupling proteins: from mitochondria to the regulation of energy balance. J Physiol 2000;529:3.

26 Shimazu T, Sudo M, Minokoshi Y, Takahashi A: Role of the hypothalamus in insulin-dependent glucose uptake in peripheral tissues. Brain Res Bull 1991;27:501.

27 Cousin B, Casteilla L, Lafontan M, Ambid L, Langin D, Berthault MF, Pénicaud L: Local sympathetic denervation of white adipose tissue in rats induces preadipocyte proliferation without noticeable changes in metabolism. Endocrinology 1993;133:2255.

28 Halberg N, Wernstedt-Asterholm I, Scherer PE: The adipocyte as an endocrine cell. Endocrinol Metab Clin North Am 2008;37:753.

29 Pénicaud L, Cousin B, Laharrague P, Leloup C, Lorsignol A, Casteilla L: Adipose tissues as part of the immune system: role of leptin and cytokines; in Kordon C (ed): Brain Somatic Cross-Talk and the Central Control of Metabolism. Berlin, Springer, 2002, p 81.

30 Cammisotto PG, Bukowiecki LJ: Mechanisms of leptin secretion from white adipocytes. Am J Physiol 2002;283:C244.

31 Ricci MR, Lee MJ, Russell CD, Wang Y, Sullivan S, Schneider SH, Brolin RE, Fried SK: Isoproterenol decreases leptin release from rat and human adipose tissue through posttranscriptional mechanisms. Am J Physiol 2005;288:E798.

32 Turtzo LC, Marx R, Lane MD: Cross-talk between sympathetic neurons and adipocytes in co-culture. Proc Natl Acad Sci USA 2001;98:12385.

33 Lee MJ, Fried SK: Integration of hormonal and nutrient signals that regulate leptin synthesis and secretion. Am J Physiol 2009;296:E1230.

34 Fu L, Isobe K, Zeng Q, Suzukawa K, Takekoshi K, Kawakami Y: β-Adrenoceptor agonists downregulate adiponectin, but upregulate adiponectin receptor-2 and tumor necrosis factor-α expression in adipocytes. Eur J Pharmacol 2007;569: 155.

35 Mohamed-Ali V, Bulmer K, Clarke D, Goodrick S, Coppack SW, Pinkney JH: β-Adrenergic regulation of proinflammatory cytokines in humans. Int J Obes Relat Metab Disord 2000;24(suppl 2):S154.

36 Vicennati V, Vottero A, Friedman C, Papanicolaou DA: Hormonal regulation of interleukin-6 production in human adipocytes. Int J Obes Relat Metab Disord 2002;26:905.

37 Géloën A, Collet AJ, Bukowiecki LJ: Role of sympathetic innervation in brown adipocyte proliferation. Am J Physiol 1992;263:R1176.

38 Klaus S, Choy L, Champigny O, Cassard-Doulcier AM, Ross S, Spiegelman B, Ricquier D: Characterization of the novel brown adipocyte cell line HIB 1B. Adrenergic pathways involved in regulation of uncoupling protein gene expression. J Cell Sci 1994;107:313.

39 Cousin B, Bascands-Viguerie N, Kassis N, Nibbelink M, Ambid L, Casteilla L, Pénicaud L: Cellular changes during cold acclimatation in adipose tissues. J Cell Physiol 1996;285.

40 Jones DD, Ramsay TG, Hausman GJ, Martin RJ: Norepinephrine inhibits rat pre-adipocyte proliferation. Int J Obes 16;349:1992.

41 Cousin B, Casteilla L, Lafontan M, Ambid L, Langin D, Berthault MF, Pénicaud L: Local sympathetic denervation of white adipose tissue in rats induces preadipocyte proliferation without noticeable changes in metabolism. Endocrinology 1993;33:2255.

42 Foster MT, Bartness TJ: Sympathetic but not sensory denervation stimulates white adipocyte proliferation. Am J Physiol 2006;291:1630.

43 Ruschka E, Elbet H, Klöting N, Boettger T, Raum K, Blüher M, Braun T: Defective peripheral nerve development is linked to abnormal architecture and metabolic activity of adipose tissue in Nsc-2 mutant mice. PLos One 2009;4:e5516.

44 Qian H, Azain MJ, Compton MM, Hartzell DL, Hausman GJ, Baile CA: Brain administration of leptin causes deletion of adipocytes by apoptosis. Endocrinology 1998;139:791.

45 Hamrick MW, Della-Fera MA, Choi YH, Hartzell D, Pennington C, Baile CA: Injections of leptin into rat ventromedial hypothalamus increase adipocyte apoptosis in peripheral fat and in bone marrow. Cell Tissue Res 2007;327:133.

46 Gullicksen PS, Della-Fera MA, Baile CA: Leptin-induced adipose apoptosis: implications for body weight regulation. Apoptosis 2003;8:327.

47 Porte D Jr, Baskin DG, Schwartz MW: Insulin signaling in the central nervous system: a critical role in metabolic homeostasis and disease from *C. elegans* to humans. Diabetes 2005;54:1264.

48 Fishman RB, Dark J: Sensory innervation of white adipose tissue. Am J Physiol 1987;29:R1630.

49 Song CK, Schwartz GJ, Bartness TJ: Anterograde transneuronal viral tract tracing reveals central sensory circuits from white adipose tissue. Am J Physiol 2009;296:R501.

50 Shi H, Song CK, Giordano A, Cinti S, Bartness TJ: Sensory or sympathetic white adipose tissue denervation differentially affects depot growth and cellularity. Am J Physiol 2005;288:R1028.

51 Shi H, Bartness TJ: White adipose tissue sensory nerve denervation mimics lipectomy-induced compensatory increases in adiposity. Am J Physiol 2005;289:R514.

52 Prunet-Marcassus B, Cousin B, Caton D, André M, Pénicaud L, Casteilla L: From heterogeneity to plasticity in adipose tissues: site-specific differences. Exp Cell Res 2006;312:727.

53 Crisan M, Yap S, Casteilla L, Chen CW, Corselli M, Park TS, Andriolo G, Sun B, Zheng B, Zhang L, Norotte C, Teng PN, Traas J, Schugar R, Deasy BM, Badylak S, Buhring HJ, Giacobino JP, Lazzari L, Huard J, Péault B: A Perivascular origin for mesenchymal stem cells in multiple human organs. Stem Cells 2008;26:2425.

54 Laharrague P, Casteilla L: The emergence of adipocytes. Endocr Dev. Basel, Karger, 2010, vol 19, pp 21–30.

Luc Pénicaud
UMR 6265, CNRS, UMR 1324 INRA, UB, 15, rue Hugues Picardet
FR–21000 Dijon (France)
Tel. +33 3 80 68 16 48, Fax +33 3 80 39 62 89
E-Mail penicaud@cict.fr

Author Index

Subject Index